THE COMING ICE AGE

LECTOR HOUSE PUBLIC DOMAIN WORKS

THE COMING ICE AGE

CHARLES AUSTIN MENDELL TABER

ISBN: 978-93-5614-293-0

Published: 1896

LECTOR HOUSE LLP
E-MAIL: lectorpublishing@gmail.com

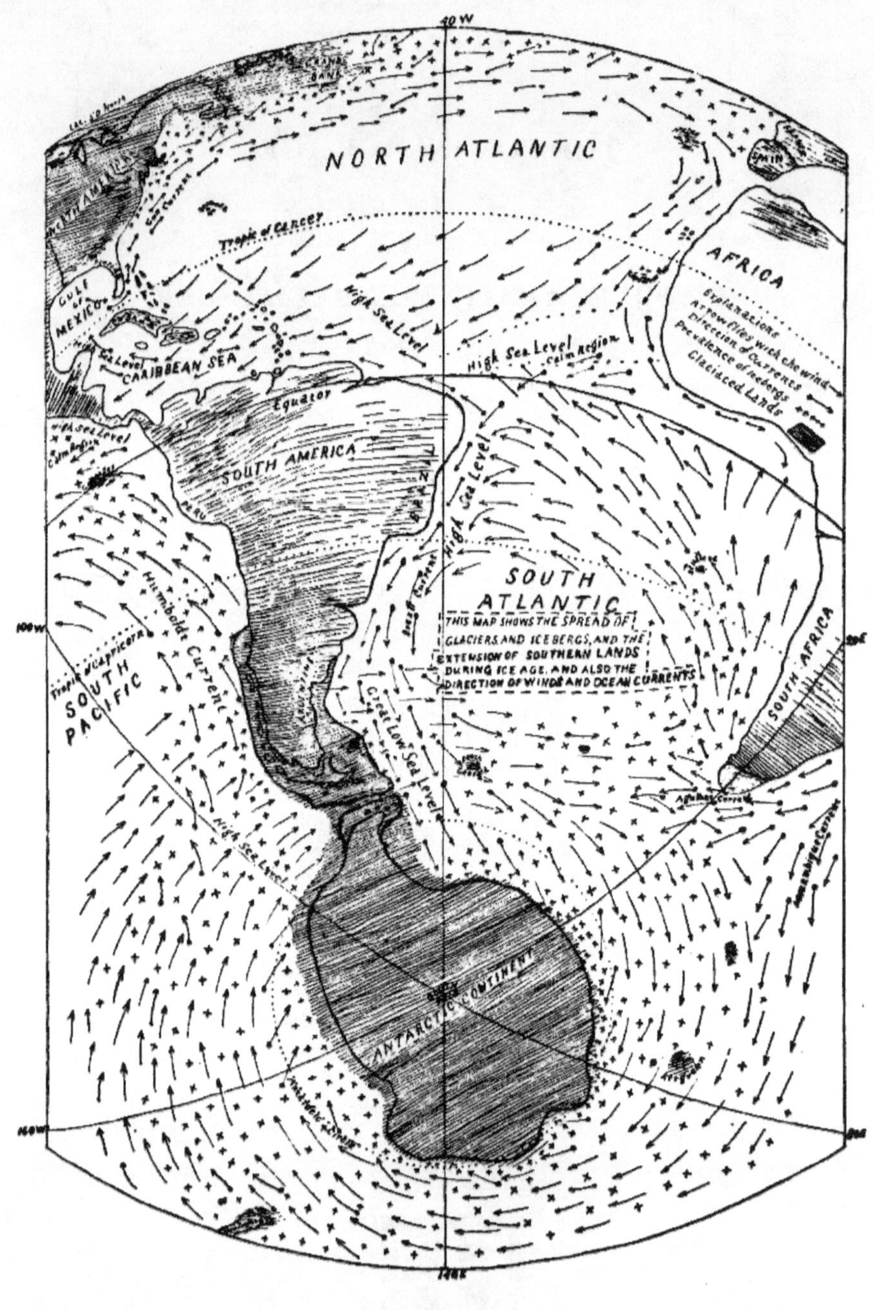

NO. 1.

THIS MAP SHOWS THE SPREAD OF GLACIERS AND ICEBERGS, AND THE EXTENSION OF SOUTHERN LANDS DURING ICE AGE AND ALSO THE DIRECTION OF WINDS AND OCEAN CURRENTS.

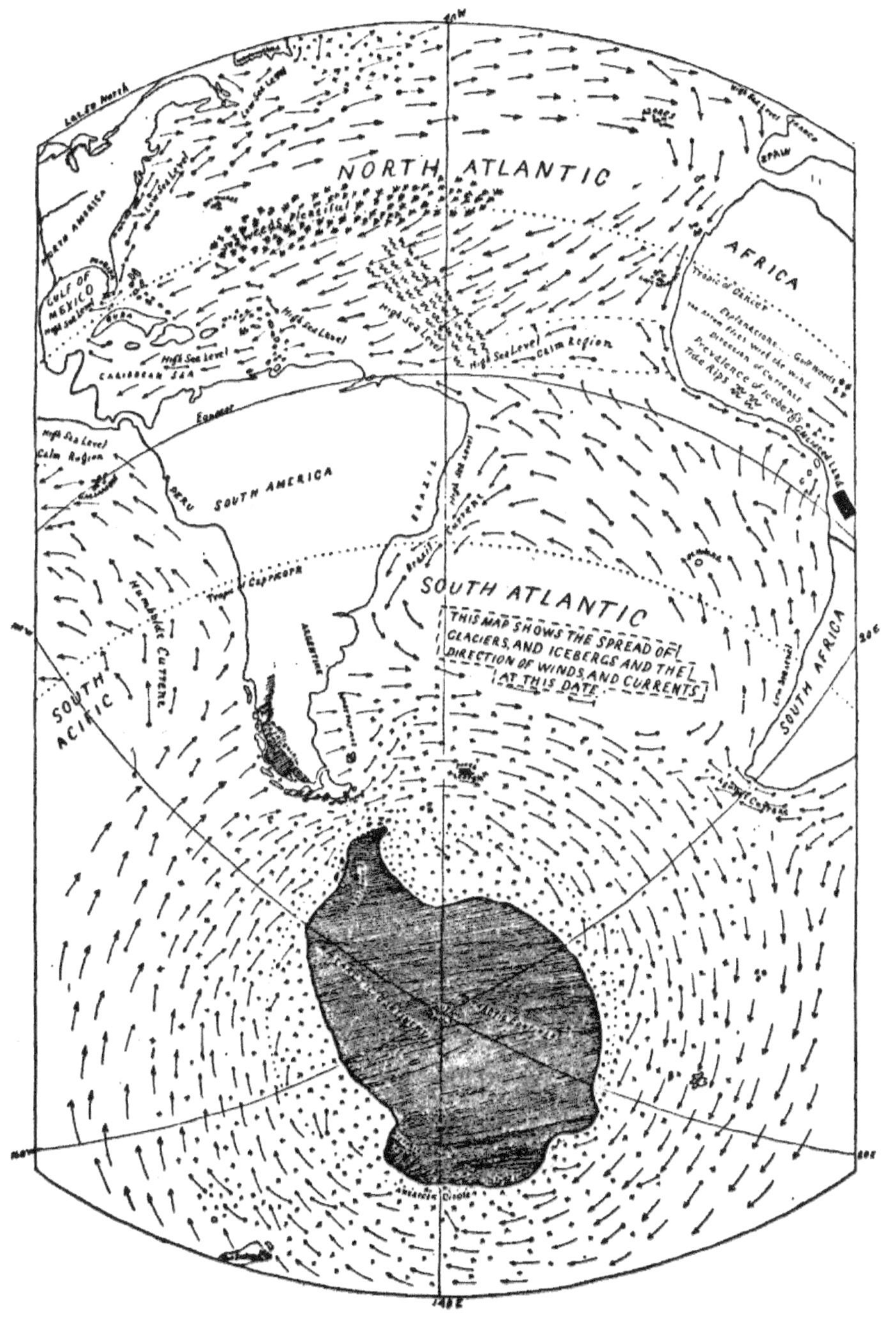

NO. 2.

**THIS MAP SHOWS THE SPREAD OF GLACIERS AND ICEBERGS
AND THE DIRECTION OF WINDS AND CURRENTS AT THIS DATE.**

THE COMING ICE AGE

BY

C. A. M. TABER

1896.

PREFACE.

THE explanations given in the following pages, in which I have sought to show the manner in which an ice age is being brought about, is an extension of a treatise on "The Cause of Warm and Frigid Periods," which I published in a small edition in 1894. And, from the small number of copies circulated, only a few came to the hands of persons particularly interested in such matter. Yet there were instances of its having proved of special interest to persons celebrated for their geological attainments, and also to instructors in physical geography. Besides, it received considerate notice in some of the leading reviews. Being thus somewhat encouraged, and thinking that the subject was too important to be neglected, I have given it further study during the last year, and meanwhile have obtained additional information from recent discoveries which has served to corroborate my views. Hence I have been able to be more explicit in my explanations in the present volume than in my earlier writings. Still, while acting as a pioneer in the matter, it will be seen that I have only attempted to expose the main outlines, as my age and failing health will not permit me to enter into the voluminous details necessary for a full explanation. In order to show why my attention has been turned to the great climatic changes which have taken place during past ages, and now threaten the future, I will repeat the introduction of my earlier publication, wherein I wrote that "the reason why I have undertaken to explain the causes which have brought about the warm and cold epochs is because of my being unable to harmonize the several theories that have been published with the general mode of action which nature pursues to-day. Having in the early part of my life been employed for a score of years in the whaling service, during which time my sea voyages were passed in cruising over the North and South Atlantic, and over the Indian Ocean, from latitudes north of the equator to the southern shores of Kerguelen Land, and along the seas of Southern Australia, I also, in my searching, cruised over the Pacific Ocean from the icy seas south of Cape Horn to the northern latitudes of Alaska, and, from New Zealand in the Western Pacific to the numerous islands in the tropical zone. And it may be said that among the chief things to be learned on such voyages was the direction of the prevailing winds and surface currents of the sea. Thus the impressions then received were in mind when, in after years, I had my attention drawn to the several theories advanced for explaining the causes which produced the warm and frigid epochs. But, so far as my marine experience goes, such theories have not harmonized with nature's mode of operating at this age of the world. Therefore, I have conceived views which, to my mind, are more agreeable to the simple operations of nature of which I have long been witness. Consequently, I have written several short essays on climatic changes since 1880, and also letters relating to the same subject, which have been published in *Science* and *Scientific American*. But the space allowed for the introduction of such matter was necessarily too limited for so wide an explanation as the subject required. The views then advanced I have again repeated, with the addition of several facts pertaining to physical geography, which, so far as I know, have never before been published."

Wakefield, Mass., U.S.A.
June, 1896.

CONTENTS

CHAPTER III.

THE SPREAD OF GLACIERS DURING COLD EPOCHS.

CHAPTER IV.

THE GLACIERS OF THE TEMPERATE ZONES.

CHAPTER V.

REMARKS ON THEORIES ADVANCED FOR EXPLAINING ICE PERIODS.

main cause of the movement of water from the northern seas at the close of glacial age; why the earth-movement hypothesis should be rejected; glaciers of Europe and Alaska; North Pacific currents; why the Pacific waters are growing cool; the lowering temperature of the northern seas; the increase of cold in Europe and Asia; falling temperature of the Andean region; General Drayson's astronomical discoveries for explaining the cause of ice periods; why the Gulf Stream was always confined to the North Atlantic; the improbability of the Indian Ocean currents entering the arctic seas; why the increase of glaciers must continue while the Cape Horn channel maintains its present capacity; comments on the coming ice age; tropical zone the abode of man during ice age; preservation of the tropical ocean fauna through the glacial

CHAPTER I.

CAUSE OF COLD AND MILD PERIODS.

Traces of ancient glaciers in temperate zones; prevailing winds the main cause of the circulation of the ocean waters between the tropical and temperate zones; general direction of prevailing winds, and how, in connection with continents, they circulate the surface waters of the sea; high and low sea-levels; separation of antarctic lands from South America; Captain Larsen's discoveries in antarctic regions; how low lands south of Cape Horn were submerged; how the winds move more surface water southward than northward; Dr. Croll's views on winds and ocean currents; under-currents of the ocean, and how caused; Gulf Stream currents; antarctic under-currents; why the winds were able to force more of the ocean waters southward than northward at the close of the Tertiary age; Mr. Alfred R. Wallace's views on Tertiary seas; how the Cape Horn channel affects the ocean currents; cause of the increase of cold in southern latitudes; how the Cape Horn channel is closed during ice age, and its effect on ocean currents and temperature of southern latitudes; the melting of glaciers from southern lands; a salt sea requisite for circulation during ice age; direction of surface currents in southern seas; Humboldt current; Agulhas current; temperature of arctic ice; movement of southern icebergs; glaciers south of Cape Horn.

It is now generally conceded by those who have given the subject much attention that the greater portion of North America above the latitude of 39° north to the shores of the Arctic Ocean has been furrowed and scoured by the action of ice.

Vast traces of ancient glaciers are also found in Europe; for it is reported that ice-sheets have left unmistakable marks of having overrun the greater part of the lands lying between the arctic seas and the latitude of the Pyrenees.

In Asia evidences of glacial action have been noticed from Northern Siberia to the mountains of Syria.

The great glaciers of Himalaya have in times past attained gigantic proportions. In Northern China huge bowlders are found scattered over the valleys, and a long distance from the mountains.

The southern hemisphere, in proportion to the extent of its land surface, shows ample traces of former ice action. From the latitude of 38° south to the southern extremity of the western continent there is said to be the clearest evidence of former glacial action in numerous bowlders scattered over the land.

On the shores of the South Pacific, from the Island of Chiloe to Cape Horn, the coast is fringed with deep fiords, which appear to be channelled out by ice, like the fiords of Norway and Greenland. And at this date the mountains of that southern region are covered with snow, and the glaciers which flow down the valleys are said to reach the tide-water as far north as the latitude of 47° south. The glaciers of New Zealand, now of Alpine proportions, during the ice age descended to the sea, and channelled the deep fiords on its south-western coast; and certain traces of glacial action have been observed in Southern Australia, and also in the province of Natal, South Africa.

Kerguelen Land is pierced with deep, narrow fiords, which have the appearance of having been the work of ancient glaciers.

The lands south of the antarctic circle are to-day supposed to be covered by an ice-sheet, of which the great ice barrier surrounding that region furnishes ample proof.

While impressed with the above reports of the work of ancient glaciers, in connection with my own observations along the shores of the several oceans, I have been led to seek for the physical causes which brought about the great climatic changes of past geological ages. And, while having the subject under consideration, I have had my attention directed to the manner in which the great prevailing winds in connection with continental lands are able to move the heated surface waters of the tropical oceans into the colder zones, and also transfer the cold waters of the higher latitudes into the tropical zones.

And it is through this grand movement of the ocean waters that we are enabled to account for the difference in the temperature of places now lying in the same parallels of latitude.

The natural methods for conveying tropical heat into the higher latitudes, and also for excluding it therefrom, are so simple and efficient that on due consideration we are able to conceive how epochs possessing mild climates have been succeeded by periods of frigidity.

It has been admitted by several writers on climatic changes that, should the tropical surface waters of the ocean be moved into the high latitudes in large volume, thus adding their warmth to the heat imparted by the sun, such combined heat would cause a mild climate. And it has been estimated that the amount of equatorial heat moved into the temperate and polar regions of the northern hemisphere by the Gulf Stream alone is equal to one-fourth of all the heat received from the sun by the North Atlantic from the tropic of Cancer to the arctic circle. Still, it appears to me, while viewing the subject from a marine standpoint, that the explainers of climatic changes have never fully comprehended the manner in which the surface waters of the ocean are moved from the tropics into the high latitudes, and returned from the high latitudes to the tropics. Consequently, they have neglected necessary and efficient natural agents in their explanatory theories, and with much learning and ingenuity have laboriously sought to show how great changes of climate could be brought about through other causes.

But when we notice the simple methods employed by nature to-day for transferring the heat of the tropics into the higher latitudes, and also the manner of excluding such heat therefrom, they appear to afford an explanation for the great changes of climate which have taken place during past ages; for it appears that the natural manner of proceeding by which heat is moved from the torrid zone into the high latitudes sufficient to cause a mild climate is through the ocean currents which are constantly set in motion by the great prevailing winds of the globe. These winds, as is well known, blow mostly from the east toward the west in the tropics, and from the west toward the east in the high latitudes.

This counter-movement of the winds, in connection with a continent extending both northward and southward from the equator over many degrees of latitude, such as obtains on the western continent, is abundantly able to create extensive depressions and elevations on the ocean's surface, and thus cause vast streams of water to move by gravity from the high sea-levels to the low sea-levels; and in this way the tropical waters have been moved during past ages, and to a considerable extent are now moved far into the northern and southern seas.

This transfer of the ocean waters is the main cause of a temperate climate being enjoyed by countries situated in the high latitudes at this age.

But, in order that the tropical currents should be able to flow into the high latitudes, in quantities sufficient to cause all lands and seas situated in such latitudes to enjoy a mild climate, it would be necessary that the land should extend unbroken, or nearly so, from the arctic to the antarctic circles. Thus, with a continent of such vast extent, the westerly winds would blow the surface waters of the ocean away from the eastern shores in the high latitudes, and so cause extensive low sea-levels; while the easterly winds of the torrid zone would heap the surface waters of the ocean against the eastern tropical shores of the continent. Consequently, the warm waters of the tropical high sea-level would be moved by gravity to the low sea-levels of the high latitudes, even to the arctic and antarctic regions, and thus afford them a mild climate. In this way we account for the mild climate enjoyed on lands and seas within the high latitudes during the warm epochs anterior to the glacial periods.

As the western continent is the only land that extends unbroken from the equator to the cold latitudes of both hemispheres, thus affording an opportunity for the prevailing winds to move the tropical waters into the high latitudes, I will call attention to that portion of the continent which extends far southward into the southern ocean, where the winds and ocean currents have the greatest range and power to affect the climate on different parts of the globe. Here we see South America separated from the antarctic continent by a wide channel of deep water, where the westerly winds blow with great force. The space now covered by this interesting channel, owing to its being situated in the high southern latitudes, must have been occupied by a channel of comparatively small capacity, or else an isthmus of low land uniting the southern portion of South America with the antarctic continent during the warm epochs when the beds of the ancient seas of the northern hemisphere contained a considerable portion of the water now swelling the southern ocean.

Therefore, the obstructions which separated the Pacific Ocean from the South Atlantic furnished opportunity for the westerly winds to force the surface waters of the sea away from the leeward side of such obstructions, causing a vast low sea-level, sufficient to attract the tropical waters heaped against Brazil by the trade winds into the southern seas in adequate quantity to cause a mild climate throughout the antarctic regions through long periods of time.

Recent discoveries have proved that these high southern latitudes have been subject to great changes of climate. According to the reports from the Dundee whalers, while searching for seal in the icy seas that surround the South Shetlands, they met with the Norwegian ship "Jason," Captain Larsen, who had traced the eastern shore of Graham Land to 68° south latitude, noting two active volcanoes.

The same mariner brought from Seymour Island fossil shells and coniferous wood of the Tertiary epoch.

These furnish sufficient evidence to show that a warmer climate once prevailed there.

At the commencement of the glacial age the obstructions which separated the South Pacific from the South Atlantic had become deeply submerged by the sea, which may have been caused by a tendency of the ocean's waters to move southward or by a comparative small movement in the earth's crust. But, on account of the stability of the crust of the earth during times so late as the glacial epochs, the submergence of this southern region was probably owing to the movement of the ocean's waters from the northern hemisphere into the southern hemisphere, which appears to have been brought about mostly through the agency of the great prevailing winds; for it seems to have happened that the prevailing winds on account of the disposition of the lands and seas were able to move more of the ocean waters southward than they moved northward during the age preceding the glacial periods. The waters thus slowly and gradually forced into the high southern latitudes must have deprived the northern hemisphere of their heaviness, and added their weight to the southern hemisphere. Therefore, the waters moved southward could not all be returned to the seas of the northern hemisphere by gravity, for the reason that the earth's centre of attraction would change in accordance with the weight of water moved from the northern hemisphere into the southern. It will thus be seen that, while the northern seas were drained or became shallow, the augmented southern oceans deeply submerged the region south of Cape Horn, thus widely separating the western continent from the antarctic lands.

Although the south-east trade winds on the eastern sides of the Atlantic and Pacific Oceans extend further northward than the north-east trade winds extend southward, owing to the heated tropical shores north of the equator being more extensive than such lands south of the equator, still, on account of the general weakness of the south-east trade winds at the equator, and also because of the obstructing northern lands, they have during remote times, and at this age, been largely prevented from impelling the surface waters of the sea into the northern latitudes in opposition to the brisk north-east trades. Furthermore, on account of the widening of the oceans as they extend southward, the surface currents setting in the latter direction have more broad and easy passages than the great currents setting northward.

Moreover, the great currents setting southward on the western sides of the oceans south of the equator are also much assisted during the southern summer months by the strong north-east monsoons which prevail along the east coast of equatorial Africa and the east coast of South America as far as the latitude of 30° south.

The South African current is impelled northward by the trade winds down the south-western coast of Africa; but it is debarred from entering the northern latitudes by the Guinea currents, and so turned away into the south equatorial current which flows into the Brazilian stream.

The Gulf Stream is much obstructed in its northern movement by the narrow Florida channel and the opposing arctic currents, and also by the trend of the North American coast eastward; while its return current on the eastern side of the Atlantic has a much less obstructed passage in its southern movement, and, while on its way past the Azores and Madeira Islands, is largely assisted by the prevailing winds.

The Brazil current, with the impelling force of a strong north-east monsoon during the summer season, has no obstruction whatever in its southern passage until it meets with an offshoot from the great drift current of the southern ocean.

And the same favorable conditions are obtained by the great currents setting southward on the western sides of the South Pacific while on their way to the low sea-levels east of Southern Australia and New Zealand. That portion of the equatorial stream of the Pacific which continues west across the Indian Ocean finds no open passage to the northern seas. Consequently, it turns south along the east coast of Africa into the southern seas.

Therefore, this current, in connection with the great currents setting southward east of Australia, offsets the great Humboldt current setting north along the coast of Peru.

In the North Pacific the Japanese current setting northward is obstructed by the narrowing of the ocean; while its return current on the American side has a constantly widening ocean on its passage southward, and also favorable winds to impel the surface waters toward the equator. Still, with all the facilities above mentioned for the movement of the ocean waters into the southern latitudes, it is probable that since the shallow seas of the northern hemisphere were drained, or much diminished, the prevailing winds have not possessed sufficient force to further augment the southern seas, because of the superior weight of the land in the northern hemisphere compared with the lands south of the equator.

It will appear to those who attribute the rotation of the earth as being the main cause of ocean currents that I am too much given over to the wind theory. But I have reason to believe, as Dr. Croll has asserted, that "the winds are the principal cause of the ocean currents, and are not due to the trade winds alone, but to the general impulse of the prevailing winds of the globe."

Dr. Croll also declares that "all of the principal currents of the globe are moving in the exact direction which they ought to move, assuming the winds to be the sole impelling cause."

Those who think that the rotation of the earth is the real cause of the movement of the great surface currents of the sea should explain in some reasonable way why the Agulhas current turns west into the Atlantic from the Mozambique stream, and why the Guinea current turns to the east from the main tropical current of the North Atlantic; for it seems that these two great currents move in direct opposition to the rotation theory, while at the same time many things go to show that they receive their motion from the winds. This view of the question will receive further attention in succeeding pages.

It is the opinion of some writers that a difference of temperature and density between the waters of the polar latitudes and the torrid zone is the principal cause of the movement of the surface waters of the ocean from the equatorial latitudes toward the polar seas, and so returned in under-currents; and this is a favorable factor for assisting the winds on some parts of the sea, especially in aiding the Brazil current in moving the surface waters from the high sea-levels abreast Brazil, and the equatorial calm belt of the Atlantic into the southern ocean, and also for favoring the surface currents setting southward on the western sides of the South Pacific and Indian Oceans.

Yet, whatever gravitating force it may possess for assisting the above-named currents, it would also act against the impelling force of the trade winds, while they were drifting the surface waters northward toward the equator on the eastern sides of the several oceans, and also to retard the returning surface currents, while being drifted by the winds southward on the eastern sides of the North Atlantic and North Pacific. Therefore, while it would seem to favor the winds in their work on the one hand, it would act as an opposing agent on other parts of the ocean. Still, the difference of temperature between the tropical and antarctic seas probably does act in opposition to the wide and brisk trade winds on the eastern sides of the great oceans south of the equator, and so prevents their impelling the surface waters northward to a great extent; and this seems to be one great cause of there being less surface water moved northward than southward over the greatest oceans of the globe.

The theory that the difference of density caused by the difference of temperature between the polar seas and the equatorial oceans made under-currents to flow from the polar latitudes, and meet in the equatorial seas, can only be carried on in the Atlantic Ocean, and in a comparatively less perfect way in the Pacific Ocean, and not at all in the Indian Ocean.

The North Atlantic being open to the Arctic Ocean, a portion of the Gulf Stream waters that enter it from the north-west of Europe do sink and return southward in under-currents; and the cold waters which pass down the east and west coast of Greenland also sink under the Gulf Stream while on their southern movement. The meeting of these arctic currents with the cold under-currents from the antarctic seas in the tropical zone is probably one cause of their cold waters rising near the surface of the sea in the torrid latitudes of the Atlantic; and the same conditions probably obtain in a somewhat less degree in the Pacific Ocean.

Yet it appears that the cold waters of the Antarctic occupy the largest space in the tropical zone, even in the North Atlantic. Dr. Carpenter, in his lectures on Ocean Currents, speaks of meeting with antarctic water so far north as the latitudes of the West India Islands; and he also says that all of the Pacific Ocean at its depths is supplied from the Antarctic Ocean, as are the cold under-waters of the tropical Indian Ocean, which extend over twenty degrees north of the equator.

Thus, from what we can learn of the antarctic under-currents, they seem to show that they are not wholly attracted northward on account of the difference of temperature between the antarctic and the tropical oceans, but partly because of more surface water being moved southward by the prevailing winds than they are able to move northward.

And it appears that, if through the winds, combined with the difference of temperature between the antarctic seas and the equatorial waters, and also because of the oceans widening toward the south, more surface water is being carried southward than northward, the waters of the under-currents so caused must rise toward the surface in the latitudes from which they were first removed. Having called attention to the fact that the prevailing winds are not able at this date to augment the southern ocean waters from the scanty northern seas, because of the preponderance of northern lands, still there is reason to believe that even now, owing to the form of continents and oceans, and the attraction of the tropical surface waters into the Antarctic Ocean because of the difference of density between the warm and cold seas, the prevailing winds of this age are able to force more of the surface waters of the sea southward than they force northward; but, owing to the superior weight of the land in the northern hemisphere, the surplus surface water forced into the southern seas is returned by gravity after being cooled by the antarctic ice, and so adding to the deep under-currents which flow with a sluggish movement over the bottom of the sea into the tropical and northern temperate latitudes. And in this way the northern oceans are maintained at their present sea-level.

The cold under-currents are probably assisted in their northern movement by whatever difference there may be in the density of the antarctic waters over the bottom waters of the equatorial seas. But, as such currents extend into the northern tropical latitudes of the northern hemisphere, it seems that the winds are the main cause of the under-currents which carry so much antarctic cold into the northern tropical seas, because the winds have forced an undue proportion of ocean surface water southward, to be attracted northward in under-currents by the preponderating northern lands.

Yet, notwithstanding the superior weight of land in the northern hemisphere, it appears that there have been periods when there was somewhat more water in the oceans of the southern hemisphere than now; for it is reported that a portion of the low lands of Australia show traces of having been submerged during late geological times.

This may have happened through an increased weight in the antarctic glaciers, which have in past ages, and probably may in future epochs, cause more of the ocean waters to be attracted southward than now obtains. But it is probable that an increase of southern ice would be largely counterbalanced by the accumulation of ice on northern lands.

Yet it appears certain that since the Tertiary epoch the waters of vast shallow seas have been moved from the northern hemisphere into the southern. The dry beds of the ancient northern seas encourage this opinion, while the comparatively small area of southern lands serves to support such views.

Still, during the ages prior to the glacial periods, while the low lands of the northern hemisphere were covered by the sea, the wide shoal channels which submerged the lower portion of North America afforded convenient passages for the surface waters of the ocean in their northern movement, and so prevented the oceans of the southern hemisphere from gaining undue preponderance.

Hence long geological ages passed away before the winds were able to force more of the ocean waters southward than they could move northward, and thus augment the southern ocean from the waters of the northern seas. But the slow growth of such immense marine deposits in the shallow seas as are found in the Florida Peninsula and other portions of that region was at length sufficient to greatly obstruct the passage of the Gulf currents in their northern movement, and thus cause conditions which enabled the winds to force more of the ocean waters southward than they could move northward after the close of the Tertiary epoch.

Mr. Alfred R. Wallace says in "Island Life" that the seas in the northern hemisphere during the Tertiary period covered a much larger area than now, and extended across Central Europe and portions of Western Asia, and the Arctic Ocean was enlarged.

As it is not likely that any portion of the waters of the sea have been absorbed by the earth during the late epochs in the world's history, therefore the ocean waters have not diminished except during cold periods, when the water evaporated from the sea was converted into ice, and, eventually, again returned to the sea.

Thus it necessarily follows that, when the seas of the northern hemisphere contained a much larger portion of the waters of the globe than at this age, the seas of the southern hemisphere must have contained proportionally less. Consequently, during such times a portion of the shoal seas of the high southern latitudes must have been dry land. Therefore, this must have been the condition of the shallow sea basins in the region of Cape Horn.

Mr. Wallace also says that "many peculiarities in the distribution of plants and some groups of animals in the southern hemisphere render it almost certain that there has sometimes been a greater extension of antarctic lands during Tertiary times."

And he also asserts that the great ocean basins have not changed, and that the form of continents has been permanent. It will thus be seen that it was through the movement of the ocean's waters southward that the low lands south of Cape Horn were covered with water previous to the frigid periods, and so caused the wide separation between the western continent and the antarctic lands.

The Cape Horn channel thus enlarged, the continuous mildness of the high southern latitudes which possessed the earlier ages came to an end, and gave place to alternate epochs of frigid and mild weather. For it appears that it is owing to the creation or enlargement of the Cape Horn channel that it is possible for frigid periods to be brought about, for the reason that its enlarged space of water prevents the westerly winds from maintaining a great low sea-level in the higher latitudes of the southern ocean; for, whenever the capacity of the Cape Horn channel is enlarged, the westerly winds, instead of maintaining a low sea-level on the South Atlantic, employ their force in impelling the surface water of the southern seas around the globe. And this work the strong westerly winds of the high southern latitudes have always accomplished whenever the Cape Horn channel was widely open, and this is what the winds are doing at this date.

Therefore, such waters of the torrid zone as are moved southward from their high sea-level, caused by the trade winds abreast the Brazilian coast, are largely turned away from the high southern latitudes. It is true, even with an enlarged Cape Horn channel, they can always flow along the South American coast to an inferior low sea-level, caused by the westerly winds blowing the surface waters of the sea away from the coast of Argentine and Patagonia; but on gaining that region they meet the cold ice-bearing currents which turn away east of Cape Horn from the great southern drift current to gain the same low sea-level which attracts the Brazil water. Consequently, the ice-bearing currents from the south, which branch off from the great southern drift current, are able to largely turn away the warm Brazil current from the higher southern latitudes; and, furthermore, the great southern drift current which passes through the Cape Horn channel, and so onward around the globe, also partly turns away the Mozambique current as well as the East Australian current, and so largely prevents their waters from warming the southern seas.

Therefore, it is evident that, whenever the Cape Horn channel obtains sufficient capacity to give an independent circulation to the southern ocean, the conditions are favorable for the increase of cold in the southern latitudes. For it is because of the large exclusion of the tropical waters from the southern seas that ice-sheets have been able to form in early periods and in later epochs on the antarctic lands, and store away the annual frosts for thousands of years, and at the same time furnish icebergs sufficient to chill the waters of the southern temperate oceans, and consequently make cold such of the surface waters of the sea as are forced into the southern latitudes by the winds in surface currents, and so returned to warmer seas in cold under-currents, and thus with such frigid combinations bring about cold periods.

Thus it appears, as I have previously shown, that it is owing partly to there being more of the surface waters of the sea forced southward by the prevailing winds than they impel northward that the cold under-currents are maintained; but it also requires an independent circulation of the southern ocean, such as I have pointed out, to cool its surface waters before they can sink and form cold under-currents.

And there is reason to believe that such cold under-currents are more efficient in lowering the temperature of the temperate and tropical oceans than even the icebergs which such under-currents move into the temperate seas. And, when it is considered that the cold antarctic under-currents fill the depths of the Pacific and Indian Oceans in the northern hemisphere, and also largely the tropical depths of the North Atlantic, I am led to believe that the frigid conditions of the ice age were concurrent in the northern and southern hemispheres. The main reasons for such belief I will explain in the following chapter.

After the foregoing explanations, showing how frigid periods are brought about through the independent circulation of the southern ocean surface waters, it is evident that, whenever through a slow natural process the Cape Horn channel is closed, a great change is wrought in the circulation of the southern ocean.

For instead of the westerly winds blowing the surface waters of the southern seas constantly around the globe, and so turning away and preventing the entrance of the tropical currents into the high southern latitudes, the strong westerly winds, whenever the Cape Horn channel is closed or greatly obstructed, would blow the surface waters away from the Atlantic side of the closed channel, and so cause a great low sea-level, sufficient to attract the ocean waters of the tropical high sea-level abreast Brazil well into the southern seas. Therefore, it is important to trace nature's slow methods of closing the wide Cape Horn channel at the perfection of an ice age.

In my previous explanations on the subject I have thought that, should the southern seas have remained at or near the same sea-level as now, through an ice period brought about in the manner I have described, ice-sheets would accumulate on the antarctic continent, and also on the southern lands of South America, sufficient to flow out into the sea and close the Cape Horn channel.

But further consideration shows the impossibility of the southern seas having maintained their present sea-level during the growth of frigid epochs which have left such ample traces of glaciers having extended widely over the lands of the high latitudes of both the northern and southern hemispheres. For it appears that the larger areas of land in the northern latitudes, embracing wide continents and large islands, must, during the growth of a frigid age, have increased the spread of glaciers many times greater in extent than could be obtained on the smaller lands of the high latitudes of the southern hemisphere.

For it is evident that the water evaporated from the sea and deposited in snow on the large continents and islands of the high northern latitudes during the growth of an ice period would, while thus diminishing the ocean waters, greatly increase the weight of northern lands. Therefore, the waters of the diminishing seas of the southern latitudes would be attracted into the northern oceans in opposition to the prevailing winds.

Thus it appears that the Cape Horn channel would be too much reduced at the perfection of an ice age to afford an independent circulation for the southern ocean, even without being filled by glaciers to the extent I have pointed out in previous essays. Still, to whatever dimensions the Cape Horn channel might be reduced at the perfection of a frigid period, the enlarged shores bordering its diminished waters would be covered by heavy glaciers that would flow into the shrunken strait, and so close it effectually. Thus the reduction of the Cape Horn channel during the advance of an ice age seems, on close consideration, to be a simple operation of nature, which in the normal course of events must have taken place.

As the closing of the Cape Horn channel has been considered by reviewers the weak and questionable point in preventing my views from gaining acceptance, it becomes necessary to be explicit concerning the manner in which the Cape Horn channel has in past ages been obstructed.

According to the charts prepared by John James Wild, the middle portion of the strait is represented as being over a thousand fathoms in depth; but, as far as I know, its true soundings have never been determined. The deep portion of the mid-channel is described as being narrow when compared with its whole breadth from Cape Horn to the antarctic continent.

And, when it is considered, with the growth of an ice age, how much of the ocean waters would be stored in the vast ice-sheets of the northern hemisphere, and consequently because of their weight a large portion of the diminished southern oceans would be attracted into the northern seas, it seems that the bottom of the shoaler waters of the Cape Horn channel, which now comprise so large a portion of its breadth, would be raised above the surface of the sea.

The one-hundred-fathom depth south of Cape Horn, now supposed to extend from longitude 70° west to 55° west, and southward to the latitude of 57°, would be a land supporting heavy glaciers for six hundred miles along the north side of the reduced channel during the advanced growth of a frigid age; and the same conditions would be obtained in the vicinity of the South Shetland. And when, in addition, we contemplate the great snow-fall of that region, and the consequent gathering of glaciers which would occur on the widened shores of the lessened channel, and the certainty of their flowing into the diminished strait, together with the immense icebergs of such an age grounding in the shoaled waters, it seems that the complete obstruction of the reduced channel would be accomplished.

While contemplating the conditions that would obtain while the Cape Horn channel was being reduced, it will be seen that the independent circulation of the icy southern ocean would be carried on to a considerable extent even after the narrowing strait was no longer able to afford space for wide drift currents, for the reason of the strong current that would be caused on account of the high ocean-level maintained by the westerly winds on the Pacific side of the diminishing channel, and the great low sea-level that would take place on its Atlantic side.

Still, as previously shown, it seems that during an advanced stage of the frigid epoch, the heavy glaciers from the enlarged northern and southern shores of the shrunken channel, together with the ponderous icebergs, blocking its waters, the closing process would at last be speedy and effective.

And on further consideration it might be said that a channel of much less width and depth would not have been of sufficient capacity to have caused ice periods so wide-spread as those that have left their traces on the continents and islands of the globe, for the reason that the independent circulation of the southern ocean would not have been sufficiently complete and long continued to have brought such world-wide cold periods to perfection.

With the Cape Horn channel closed, as above explained, there would be, as I have asserted, a great change wrought in the circulation of the southern ocean; for instead of the westerly winds blowing its surface waters constantly around the globe, and so turning away and preventing the entrance of tropical currents into the higher latitudes, the strong prevailing westerly winds would blow the surface waters of the sea from the Atlantic side of the closed Cape Horn channel, and so cause a great low sea-level, sufficient to attract the ocean waters of the tropical high sea-level abreast Brazil well into the southern seas.

The winds of the southern westerly wind-belt being stronger in that region than on any other portion of the globe, consequently they are able to do nearly as much work while drifting surface water as the belt of westerly wind of greater width on other parts of the southern seas. Thus a person who has had a long experience with the forcible westerly winds of the southern ocean can well understand their ability for disturbing the ocean waters in the latitudes of the Cape Horn channel.

The drift currents of this region are moved by the winds and waves from one to four miles an hour. Therefore, with the Cape Horn channel closed, there is nothing more certain than that the westerly winds would be able to cause a vast low sea-level on the Atlantic side of the closed Cape Horn strait, and that the waters of the high tropical sea-level abreast Brazil would be attracted to its wide depression, as shown on map No. 1.

The tropical waters thus attracted far southward would be cooler than the tropical waters of to-day, owing to the great amount of cold imparted to the ocean by the numerous icebergs of a frigid age. Still, they would begin the slow process of raising the temperature of the southern ocean, and would in time carry sufficient heat into the southern regions to melt the ice from all southern lands; for, in addition to the Brazil currents, the waters of the high sea-level of the tropical Indian Ocean which pass southward down the Mozambique channel would reach a much higher latitude than during periods when the Cape Horn channel was open.

The ice periods of the northern and southern hemispheres being concurrent, a condition which I shall explain in another chapter, makes it obvious that during the melting of the glaciers from the antarctic continent and other southern lands the depleted Cape Horn channel could not gain sufficient capacity to give an independent circulation to the southern ocean during the melting of the southern ice-sheets, on account of the diminishing heaviness of the antarctic ice and the greater weight of the extensive glaciers and augmented seas of the northern latitudes. Consequently, it seems that the southern seas would continue in a lessened state while the glaciers were being melted from the northern hemisphere, as was the case during the melting of the ice from the southern hemisphere; and, furthermore, during such times the glaciers which overrun all the low lands and shoal waters of the Cape Horn region would, on account of their position being to the windward of the tropical currents, be the last great mass of ice to melt from the southern hemisphere.

Therefore, it seems that the Cape Horn channel would continue closed or greatly obstructed while the glaciers were being melted from the lands of both hemispheres. Thus at length a mild climate would extend over the globe, and so remain until the prevailing winds slowly forced the surface waters of the sea into the southern ocean in the manner explained in previous pages, thus filling the Cape Horn channel to its present capacity, and again restoring the independent circulation of the southern ocean.

While contemplating the conditions that would obtain during the melting of the ice from the antarctic lands, it will be seen that the tropical waters attracted to the great low sea-level to the leeward of the closed Cape Horn channel would eventually enter the great bight of the antarctic continent to the eastward of Graham Land, where Captain Weddell sailed to the latitude of 74° south. This deep gulf, owing to its situation, would receive the full impact of the southern movement of the tropical currents; and, as the warm waters spread over the wide sea-level, the westerly winds would convert them into a drift current, and under such conditions would be driven along the shores of the antarctic continent, past the South Indian and Pacific Oceans, and eventually, after undergoing a cooling process from the long icy passage, be forced against the Pacific side of the closed Cape Horn channel and the western Patagonian coast.

While regarding the circulation of the sea during an ice age, it may be said that the ocean's being composed of brine was the cause of its waters being able to circulate in frigid latitudes where fresh water would congeal. Consequently, this is one of the reasons why successive periods of frigidity and mildness have been brought about; for with an ocean of fresh water, repeated epochs of cold and warmth could not have occurred, because a sea composed of fresh water would have congealed while circulating in the high latitudes during a frigid age. Therefore, it required a sea of brine to maintain a liquid state during the low temperature of an ice period.

For, while the cold of a glacial age increased, the saltness of the sea increased also, because of the great amount of fresh water evaporated from the ocean, and stored in ice-sheets on the great continents and islands of the globe. Thus the briny sea was maintained in a liquid state, while washing vast ice-fields and glaciated shores and floating the numerous icebergs of a freezing age. The cold which radiated from such ice-bound seas must have been severe; but meanwhile the evaporation from the ocean was much reduced, while the saltness and coldness of the sea increased, and so prevented the ice of a glacial period from gaining invincible proportions before the independent circulation of the southern ocean was arrested. Therefore, the remaining warmth of the tropical waters after gaining free access to the antarctic latitudes was able to overcome the accumulated cold of that frigid region.

At this date the observant navigators who have visited the antarctic seas report that the surface currents above the latitude of Cape Horn, while being drifted eastward by the prevailing westerly winds, also set toward the antarctic ice cliffs, as shown on map No. 2.

The reason why this southerly set of the surface currents becomes noticeable above the latitude of 55° south is because the tropical currents which set southward from the torrid latitudes on the western sides of the Atlantic, Indian, and Pacific Oceans, although largely turned away from the high latitudes by the westerly winds and drift currents, are also able to send sufficient water into the great belt of westerly winds to furnish water for the deep under-currents setting northward from the antarctic shores. Thus the surface waters moving from the north in order to gain the higher latitudes, after entering the westerly wind-belt, are moved in drift currents by the impelling winds easterly over many degrees of longitude, and also at the same time slowly southward among the cooling icebergs, because of the attraction caused by the difference of temperature and density between the northern drift waters and the icy seas of the antarctic ice barrier. Consequently, the gradual movement of the surface waters of the westerly wind-belt southward before entering the higher latitudes is not generally apparent; for it is after they enter latitudes where the globe becomes much reduced in circumference that their southern movement in the contracted seas becomes more noticeable. The impact of this southerly current, which finds its outlet in deep under-currents, and retards somewhat the increase of ice on the southern continent at this date, also largely prevents the small icebergs and field-ice from floating northward, away from the antarctic ice barrier; for it is such large icebergs as penetrate the deep under-currents that are the best able to move into the more temperate latitudes.

From the above explanations it will be seen that the impact of surface water against the antarctic ice barrier when the Cape Horn channel was closed would greatly assist the tropical waters attracted to the great low sea-level to the leeward of the obstructed strait to wash the antarctic shores while being drifted eastward by the westerly winds over the southern ocean against the Patagonian coast and the Pacific side of the closed channel, and there causing a high sea-level. This movement of the winds and currents encircling the antarctic continent is shown on map No. 1.

The vast, high sea-level caused by the westerly winds drifting the surface waters against the Patagonian coast would obtain a much higher plain, were it not that so much of the water of the great drift current was required to feed the antarctic under-current which constantly sets northward from the antarctic shores; yet it would be sufficient to greatly increase the volume of the Humboldt current, which would flow in the same direction it now flows, down the South American coast to the equatorial latitudes, where it would become the main source of the great equatorial stream, and thus offset the increased southward flow of the equatorial waters through the Brazil and Mozambique streams.

The equatorial stream, with its increased volume, would also move, as it moves to-day, across the Pacific; and, on gaining the western side, after sending off large streams to the northern and southern latitudes, it would pass through the East India passages into the Indian Ocean, where it would be drifted westward by the trade winds and cause a high sea-level abreast the east coast of Africa, and so become the source of the great Mozambique current, which would flow southward along the east coast of Africa, and, with the Cape Horn channel closed, would gain a much higher latitude than it would with the channel open. At this age, when the continuation of this great equatorial stream gains the latitude of the Cape of Good Hope, its waters are largely turned eastward by the great drift current of the southern ocean.

Still, a considerable portion of its waters turns toward the west, forming the Agulhas current, which flows around the Cape of Good Hope into the Atlantic, where it mingles with the cooler currents which branch off from the great southern drift current; and so, in connection with the latter, it is attracted to the low sea-level caused by the south-east trade winds abreast the south-western coast of Africa, and from thence moved as a drift current by the trade winds to the equatorial Atlantic and coast of Brazil. Thus it will be seen that the Agulhas current, even with the Cape Horn channel in possession of its present wide capacity, serves to retard somewhat the advance of a cold period.

The Agulhas current at this date also partly serves to replenish the water which is forced from the South Atlantic by strong westerly winds into the Southern Indian and Southern Pacific Oceans. For it appears that more water is now removed by such winds from the South Atlantic than enters it from the South Pacific, even through the enlarged Cape Horn channel of this date; and this fact seems to favor an impression that a portion of this enlarged channel existed prior to the glacial periods, but with its waters so much reduced as to be unable to give the southern ocean an independent circulation sufficient to exclude the tropical currents from reaching the high southern latitudes in adequate volume to maintain a mild climate in the southern hemisphere.

For previous to the glacial age, with little or no ice gathered on the antarctic lands, it seems that a strait possessing one-half the capacity of the Cape Horn channel of the present age could not prevent the Brazil current and the Agulhas stream from flowing into the southern ocean in quantities sufficient to make it impossible for glaciers to form on southern lands.

Thus it is probable that a reduced channel separated the western continent from the antarctic lands even in the mild eras previous to the glacial epochs.

The Cape Horn channel, at the present age, with a capacity sufficient to largely maintain an independent circulation for the southern ocean, is still only one-third of the breadth of the westerly wind-belt of the southern seas. Therefore, the drift currents do not all pass through it from the Pacific into the Atlantic. Consequently, a considerable portion of the drifted water turns northward west of Cape Horn, and so forms the Humboldt current.

The Agulhas stream, which even now assists in replenishing the South Atlantic with tropical water, would, during the perfection of a glacial period, with the Cape channel closed, be a much stronger stream than it now obtains with the Cape channel possessing its present enlarged capacity, for the reason that the South Atlantic waters would continue as now to be forced eastward by the westerly winds, while they could not be replenished, as they are to-day, directly from the South Pacific.

Consequently, the waters of the South Atlantic Ocean would be correspondingly reduced.

Such conditions alone would greatly increase the volume of the Agulhas stream at the culmination of a frigid age. Therefore, the work of subduing a frigid period in the southern hemisphere after the Cape Horn channel was closed would not rest on the Brazil current alone, but also on the great equatorial stream of the Pacific and Indian Oceans.

Yet during such frigid times the sources of the equatorial stream would be greatly chilled by its two great feeders, the Humboldt current and the returning Japanese current, both of which flow down from the high latitudes and meet in the equatorial latitudes on the eastern side of the Pacific, thus cooling the source of the great equatorial current.

But this latter stream, while on its long western passage across the Pacific and Indian Oceans, beneath a torrid sun, with only one cold feeder from the south which approaches it along the west side of Australia, would, on its long tropical journey, be able to obtain considerable warmth, even during an ice period, to supply the Mozambique and Agulhas streams, and so greatly assist the Atlantic waters in bringing about a mild period. Still, the process of subduing the cold of the southern latitudes would be slow, even with the Cape Horn channel closed, because of the vast collection of ice burdening the sea and land.

Yet there were conditions that were naturally brought about to favor the process of returning warmth; for it appears that, when the southern ocean was made shallow because of a considerable portion of its waters having been moved into the northern hemisphere, it will be seen that the conditions were more favorable for the westerly winds to create drift currents than would be the case on deeper seas. Therefore, the high and low sea-levels caused by such winds would be greater on a shallow ocean than would occur on deeper waters. Thus the low sea-levels of the shallow southern sea would have strong attraction for tropical surface waters, and so increase the thickness of its warm drift currents, and at the same time its lessened depths would have less capacity for the storage of cold water to reduce the temperature of the under-waters of the tropical zone.

And, furthermore, when the southern ocean was shallow, New Zealand acquired a longer extension of land to the north and south. Consequently, the enlarged low sea-level on its eastern side attracted more tropical water into the southern latitudes than now.

So, according to the conditions I have pointed out, the ice-sheets would at length melt away, and a long period of mildness would succeed on account of the length of time it would require after the ice disappeared from the earth for the prevailing winds to move the surface waters of the augmented northern seas into the southern ocean, and again restore its independent circulation, and so, after a considerable lapse of time, bring about the geographical and climatic conditions existing at the present date, which can be seen on map No. 2, which shows that a cold period has already made considerable advance in the southern hemisphere, the southern continent and islands being covered with glaciers, and the prevalence of icebergs as far north as the latitude of 35° south.

Moreover, when we consider that the independent circulation of the southern ocean is caused by the westerly winds blowing its surface waters constantly around the globe through the open Cape Horn channel, and so largely preventing the tropical currents from entering the high southern latitudes, and how, in consequence, the cold is slowly on the increase through the constant accumulation of ice on the lands and in seas of the southern latitudes, it appears that a frigid age is slowly progressing in the southern hemisphere. For it seems that continental ice-sheets should not only be able to retain their freezing temperature, but also the mean of the low temperature in which they were formed, for a considerable length of time, and so impart their extreme coldness in the shape of icebergs into such seas as border on the glaciated lands.

It has been proved at Point Barrow that strata of ice and gravel can maintain a wintry temperature through the summer months. Captain G. B. Borden, keeper of the refuge station in that region, states that Lieutenant Ray, of the Signal Service, excavated through ice and gravel to a depth of forty-one feet, and that the lower portion of the excavation maintains a temperature 15° Fahrenheit above zero the year around. Therefore, with the probability of southern glaciers obtaining a temperature of over 15° Fahrenheit below the freezing point, we can well realize the frigidity imparted to the southern oceans while melting numerous immense icebergs, and consequently will conclude that the temperature of the southern latitudes is gradually lowering.

The icebergs of the antarctic seas would not move northward into the temperature latitudes so readily as they now do, were it not that the general southward set of the southern ocean currents were interrupted by the movement of northerly surface currents in the longitudes of the low sea-levels, caused by the westerly winds drifting the surface waters of the sea from the eastern coasts of Southern South America and New Zealand. For it is owing to the low sea-levels thus created, in connection with the deep under-currents which set northward from the ice cliffs of the antarctic lands, that many icebergs are enabled to move into the temperate latitudes, especially to seas north-east of the Falkland Islands.

On other portions of the southern ocean above the latitude of 55° south the surface waters, while being drifted eastward by the strong westerly winds, also set toward the antarctic shores, and so furnish water for the cold under-currents which set northward from that frigid region. Thus from such parts of the coast only the largest bergs, which require a deep sea to float them, are moved by the under-currents into the temperate latitudes. Therefore, it happens that, while an ice period progresses, and the antarctic icebergs increase in size, the more readily the cold, deep under-currents force them into the temperate zone, in opposition to the winds and surface currents.

The icebergs, after gaining the temperate latitudes, are moved more or less eastward by the westerly winds and drift currents, and so are scattered over the southern temperate oceans, where the melting bergs impart whatever coldness they were able to store up while forming in the antarctic regions.

The low sea-levels caused by the westerly winds to the leeward of New Zealand and to the leeward of Argentine, not only cause the ice-bearing currents to set northward, but they also cause the tropical currents to make considerable inroads into the high southern latitudes. This is the reason why the lands are less burdened with ice on the antarctic shores opposite Cape Horn than on other parts of that glaciated continent.

The tropical currents which turn southward east of New Zealand largely mingle their waters with the great southern drift current, and so are carried through the Cape Horn channel. Owing to this cause, the antarctic lands abreast Cape Horn are less burdened with ice than other portions of the antarctic shores.

Thus, were it not for this penetration of warm waters southward, the antarctic coasts south of Cape Horn, because of the great snow-fall of that region, would obtain heavier glaciers than other portions of the southern continent. But the time is slowly coming when, with a lower temperature, the ice-sheets on the lands in the vicinity of the South Shetlands will attain greater thickness than the glaciers on other shores of the antarctic continent.

Hence it appears that, when the several agents for producing and distributing cold in the southern latitudes are taken into consideration, the immense and continuous storage of ice on the southern lands, which adds to the wide-spread fleet of icebergs that float the southern temperate seas, and also the vast movement of cold antarctic water into the temperate and tropical oceans in deep under-currents, combined with the increasing coldness of the westerly winds, are now slowly bringing about in the southern hemisphere a period of frigidity.

CHAPTER II.

HOW ICE PERIODS IN THE NORTHERN HEMISPHERE ARE BROUGHT ABOUT.

Northern seas during Tertiary age; Gulf Stream during Tertiary times; the origin of a cold period in the northern hemisphere; remarks on Gulf Stream and arctic currents; circulation of arctic waters; arctic channels during ice age; how the weight of glaciers in the northern hemisphere attracts the waters of the southern seas during ice age; Professor Prestwich on the submergence of European lands; the great Atlantic tide rips the head-waters of the Gulf Stream; high sea-level of Atlantic calm region; tropical Atlantic currents; Sargasso Sea; arctic and Gulf Stream currents; Pacific Ocean currents; slow growth of an ice period; reduction of Cape Horn channel; permanence of antarctic glaciers elevated above the snow-line during mild periods.

A LARGE number of geologists are of the opinion that during the whole of the Tertiary period the climate of the northern temperate and arctic latitudes was uniformly warm, without a trace of intervening frigid periods. I have before explained why the climate was made warm in the southern hemisphere during the Tertiary epoch, and how on the closing of that age, and subsequently, a considerable portion of the ocean waters had moved from the northern hemisphere into the southern.

Therefore, the northern seas during Tertiary times covered a much larger area than have obtained during periods following that mild epoch. So, when the low lands of Europe were submerged, the Baltic, Caspian, and other neighboring seas, now land-locked, were a portion of an enlarged Atlantic. Consequently, the westerly winds blew over a much wider North Atlantic than during the later periods.

Thus the high sea-level caused by such winds on its European side was greater than has since been obtained with the Atlantic of less breadth. This high sea-level, composed largely of drift water from the ancient Gulf Stream, had convenient access to the enlarged Arctic Ocean, which then covered the low plains of Northern Europe and Siberia. And owing to the trend of elevated lands north-eastward, which then formed the southern shores of the Arctic Ocean in those regions, the warm waters of the high sea-level of the Eastern North Atlantic found an easy passage into the arctic seas; for, while they moved over the European and Siberian seas to the north-east, they had the assistance of the westerly winds well into the arctic seas, from which position they were attracted across the Arctic Ocean to the low sea-level abreast Labrador and Davis Strait.

The Gulf Stream of Tertiary times comprised a much larger area than it now obtains; for with Florida and a large portion of the Gulf States submerged, and a wide, shallow sea covering the Mississippi valley and the Great Lake region, the tropical waters of the enlarged Gulf of Mexico moved from their vast high sea-level to the low sea-level abreast British America and Labrador, without being confined to the narrow Florida channel. Thus with an enlarged Gulf Stream in possession of a wide and clear passage leading northward, in connection with a mild period in the southern hemisphere, giving warmth to the southern oceans, the resources of the ancient Gulf currents for warming the northern regions were so ample and inexhaustive they were fully able to maintain a mild climate on the shores of the European seas, and also on the shores bordering the Arctic Ocean, during the Tertiary epoch.

Furthermore, the Humboldt current, which had its rise in the mild southern seas of that age, mingled its warmth with the equatorial current of the Pacific, which in turn gave its warmth to the Japanese current. Therefore, the latter stream under such conditions was competent to maintain a mild climate on the North Pacific coasts.

The origin of a cold period in the northern hemisphere was largely owing to the changed condition of the northern oceans following the close of the Tertiary epoch. The movement of the ocean waters into the southern hemisphere lessened the area of the Arctic and North Atlantic Oceans, and brought them to their present reduced limits, and also diminished the volume of the Gulf currents.

This great geographical change, in connection with a cold period progressing in the southern hemisphere, and so increasing the coldness of the Japanese current, and the cold antarctic currents, previously explained, which set northward on the bottom of the sea through the torrid latitudes even into the North Pacific and North Atlantic Oceans, were altogether sufficient to cause conditions favorable for the advancement of a cold period in northern latitudes. Besides, with reduced northern oceans and a diminished Gulf current, conditions were favorable for an independent circulation of the arctic waters, such as is being carried out at the present time. Hence an explanation of the movements of the ocean waters of to-day will explain the conditions which caused the northern ice periods in times past, as well as those to come in a future age. Although the conditions are such that the independent circulation of the arctic waters cannot be so well performed as the independent circulation of the southern ocean, still the open arctic channels are able to prevent the tropical Gulf Stream water from largely entering the higher northern latitudes. For it is certain that the prevailing westerly winds blow the surface waters of the North Atlantic away from the eastern shores of North America from Georgia to Labrador.

Consequently, the low sea-level thus caused attracts the waters of the Arctic Ocean southward through Baffin's Bay and Davis Strait, and likewise down the east coast of Greenland, thus surrounding that large island with an arctic temperature, and so causing it to become a land of glaciers, which are constantly launching icebergs into the sea to cool the waters of the northern oceans. The tropical waters of the high sea-level of the Gulf of Mexico also seek the low sea-level abreast the American coast, thus causing the Gulf Stream. This great ocean current, being the main conveyer of tropical heat into the high latitudes of the North Atlantic, calls for particular notice. The great gravity currents, of which the Gulf Stream is one of the most conspicuous, are moved by small gradients.

Hence the gradient which causes the Gulf Stream waters to move out of the Florida passage is small. The levellings which have been made place the surface waters of the Gulf of Mexico as being about one metre higher than the Atlantic abreast New York, the pressure of the higher Gulf waters toward the low level of the Atlantic being nearly equal in the narrow Florida channel from the surface to the bottom of the stream. Therefore, according to descriptions given by Commander Bartlett, the warm stream moves like a river over the hard level floor of the channel; but to the northward of the Bahamas, abreast Cape Hatteras, the stream spreads out in fanlike form, and flows over a bed of cold water of great depth.

A bed of cold water is found to cover the bottom of all the deep oceans that are accessible to the antarctic seas, through which the cold water is mostly supplied, as I have before pointed out.

But the cold water which underruns the Gulf Stream is probably furnished by the arctic waters which move down Davis Strait and the east coast of Greenland. The Gulf Stream, as it widens and becomes more shallow, is, through its exposure to the westerly winds, gradually converted into a drift current; and in this way its surface waters are forced over abreast the shores of Western Europe, where it imparts its warmth to a wide region, and also causes a high sea-level. A portion of the waters of this high sea-level turn southward to replenish the waters which have been moved by the trade winds from the eastern tropical North Atlantic over into the Caribbean Sea and Gulf of Mexico, while its northern and smaller portion mingles with the Arctic Ocean waters north of Europe. These latter waters, having escaped from the westerly wind-belt, and acquired a high sea-level, and also made cool on mingling with the icy arctic seas, lose a part of their bulk on becoming chilled by sinking and returning in under-currents to the seas from which they were forced by the south-westerly winds; while the larger remaining surface waters set across the Arctic Ocean over to the northern coast of Greenland, and so down the east and west coasts of that large island to the low sea-level abreast the American coast, where the cold waters not only crowd the Gulf Stream from the shore, but they also sink under it, and form the vast bed of cold water over which the Gulf currents flow. This cold underflow of water southward probably joins the deep antarctic currents south and south-east of the Bermuda Islands, and returns to the tropical latitudes a portion of the water that is carried into the Arctic Ocean by the Gulf Stream.

There are times during the late summer and early fall months when the arctic channels are considerably obstructed by icebergs, and the low sea-level of Davis Strait and Baffin's Bay, with the assistance of occasional south-east winds, is able to attract the temperate waters of the Atlantic as far north as the Arctic Circle. Also from the same cause the icy waters which flow down the east coast of Greenland are attracted along its southern and south-western shores into Davis Strait.

Yet at the same time the icy waters which flow from Smith's Sound and other arctic channels move in a counter-current down the westerly side of Baffin's Bay and Davis Strait, and so carry the icebergs and field-ice past Labrador and Newfoundland well on to the borders of the Gulf Stream. And, according to Lieutenant Maury, the westerly gales of the winter months force the temperate waters of the Atlantic, which pertain to the Gulf Stream, several degrees away from the south-east coast of Greenland. Therefore, during such seasons the surface waters of the returned arctic currents, which flow down the east coast of Greenland and Davis Strait, are drifted past Southern Greenland and Iceland, and so onward into the arctic seas, north of Europe. Thus the arctic waters maintain an independent circulation sufficient to largely exclude the Gulf Stream from the arctic seas, and surround Greenland with an arctic temperature; and it is on this account glaciers have formed on Greenland and other arctic shores, and such glaciers are probably increasing, as every iceberg launched from the frigid lands and floated to the lower latitudes lowers somewhat the temperature of the North Atlantic, and so causes conditions favorable for larger accumulations of ice on the arctic shores.

Yet it is probable that an ice period extending over the northern temperate zone could not be perfected by this process alone, should the tropical and southern oceans maintain their present temperature. But, with the assistance of a frigid period in the southern hemisphere to cool the ocean waters, and thus lower the temperature of all tropical currents, including the Gulf Stream and Japan currents, an ice age could be brought about in the northern hemisphere equal in intensity to the glacial periods of the past.

And, when we know that a considerable portion of the heat carried into the northern latitudes by tropical streams is largely derived through the mingling of the waters of such currents with the warm waters of the southern tropical oceans, it is evident that the ice periods of the northern and southern hemispheres were concurrent; although the culmination of the northern frigid period would be somewhat later than the perfected southern ice age, on account of the northern seas requiring the assistance of the cold oceans of the southern hemisphere to perfect a northern ice age.

The small area of the northern seas, compared with the southern oceans, and the wide mingling of the ocean waters of the hemispheres, make it evident that the comparatively scanty northern seas could not bring about or maintain either a frigid or mild period in opposition to the superior oceans of the southern hemisphere.

On the consummation of an ice period in the northern hemisphere heavy glaciers covered the larger portion of its continents and islands, which added so much weight to the northern lands as to attract the waters of the southern oceans into the northern latitudes, as I have before explained.

Thus, when the ice was mostly melted from the lands of the southern hemisphere, the heavy ice-sheets that remained on the extensive northern lands would still continue to attract the warm waters of the southern seas into the northern oceans; and in this way the Japanese and Gulf currents would gain a higher temperature and greater volume, and thus add to their ability for melting the northern glaciers wherever they were able to flow, and so hasten the growth of a mild era in the northern hemisphere.

And it seems reasonable to suppose that there was more water in the northern hemisphere on the ending of its ice period than at this age; yet it appears that it was returned to the southern hemisphere during a short period by the prevailing winds in the manner which I have previously explained.

Therefore, there are but few traces of such flowage to be found in the glacial drift, especially with the scarcity of marine life after the rigor of a frigid age.

An article in *Science*, July 5, 1895, written by Agnes Crane, states that Professor Joseph Prestwich has recently contributed a suggestive memoir on this subject to the Philosophical Transactions of the Royal Society. It treats of the evidence of a submergence of Western Europe and the Mediterranean coasts at the close of the glacial period; and in a previous paper communicated to the Geological Society of London, in 1892, the author gave evidence, deduced from personal observation, of the submergence of the south of England not less than a thousand feet, at the close of the glacial epoch.

Since that time the flood of water which flowed all of the low lands of the high northern latitudes has been returned to the southern seas, because of the force of the prevailing winds in connection with the great oceans which open so widely toward the south, the force of the winds being assisted through the attraction caused by the difference of temperature in the surface waters of the vast southern temperate oceans and the antarctic seas, and in this manner bringing about the geographical conditions of to-day which favor the return of another ice age.

It is said by those who attribute the great currents of the ocean to the rotation of the earth that the winds have little to do in causing such currents as the Gulf Stream. But my impression is that the southern portion of the Gulf Stream waters, after being drifted by westerly winds over abreast Europe, are attracted to the low sea-level in the vicinity of the Canary Islands, to be moved by the trade winds toward the equatorial calm belt and the West India Islands. And during my many months' cruising over these seas I have had my attention directed to the singular action of the surface waters, while being impelled by the trade winds toward the West India sea; for during the first fifteen hundred miles of their passage they are moved by the prevailing easterly winds without much apparent resistance or unusual disturbance. But on nearing the longitude of Cape St. Roque, and having acquired a high sea-level from which there is no easy or wide outlet, the impelled surface waters begin to rebel against the forceful winds, and cause a remarkable commotion in the shape of tide-rips and white-capped ripples, which extend from the equator in a northerly direction to the latitude of about 19° north, thus crossing the central portion of the north-east trade-wind belt, with a breadth of over three hundred miles, as shown on map No. 2.

This disturbed region where the winds and waters conflict is the probable fountain-head of the Gulf Stream. The reason why the surface waters of this disturbed portion of the Atlantic do not flow peacefully along through the West India passages into the Caribbean Sea and Gulf of Mexico is because of their narrow outlet at the Florida channel. For it is mainly through this narrow channel that the vast waters of the tropical high sea-level are attracted to the low ocean-level of the Western North Atlantic.

Thus it seems that the great fountain-head of the Gulf Stream is situated between the wide tide-rips and the Caribbean Islands. The waters from this high ocean-level enter the Caribbean Sea mainly through the several passages south of Guadeloupe; while the northern portion of the raised waters set mostly toward the north-west, and so unite with the eastern portion of the Gulf currents after they enter the Atlantic. Still, the great high sea-level which presses against the Windward Islands, being somewhat higher than the Caribbean Sea, forces its waters through the island passages in quantities sufficient to supply the Gulf Stream; and there are times when the winds are so strong and favorable that all of the passages east of Cuba conduct water into the Caribbean Sea, the cold under-waters entering the deeper channels as well as the warm surface waters. Yet the currents setting through these numerous channels are subject to fluctuations, and so also is the Gulf Stream which they supply.

That portion of the high sea-level south of Guadeloupe receives considerable assistance as a feeder for the Gulf Stream through being connected on the south by the great high sea-level abreast Brazil and the great high sea-level of the equatorial calm belt. The latter high level is caused by the trade winds, which generally blow briskly down the coast of Sahara, and also further off shore, and ending south of the Cape Verde Islands somewhat abruptly in the equatorial calm belt.

The south-east trades which blow over the Eastern and Middle South Atlantic terminate on the southern side of the calm region. Therefore, the two trade winds impel the surface waters of the tropical Atlantic from opposite directions directly toward the calm belt, and so raise its waters above the common level of the sea.

This is the opinion of the writers of the South Atlantic Directory. Still, it is probable that the high ocean-level of the calm belt is but slightly raised above the common level of the sea, on account of the trade winds having to contend against the tendency of the warm tropical surface waters to move toward the polar latitudes. The calm belt expanse which extends from Africa, where it attains its greatest width, gradually narrows as it extends westward to the longitude of Cape St. Roque, where it attains its highest sea-level, on account of the borders of its narrowing space being impelled westward by the trade winds.

The movement of the waters of this high ocean-level is mostly toward the west, forming a portion of the equatorial current of the Atlantic. The reason of its western movement is on account of its raised waters being able to supply a portion of the Gulf Stream with water which is sent off in a westerly current along the South American coast, west of Cape St. Roque into the Caribbean Sea; while, on the other hand, it joins with the great high sea-level abreast Brazil, and so unites with its great southern current. The gradient of the high sea-level of the calm belt on its southern side probably extends south of the equator, on account of the south-east trades being weak in latitudes near the equator; while on the north side the north-east trades generally blow brisk and end more abruptly, so producing a gradient of less width than that of the South Atlantic side.

It does not appear that the seas of the high northern latitudes gain an undue proportion of the tropical Atlantic waters, because of the south-east trades extending north of the equator, on account of such winds being weak, and the waters of the high sea-level of the Western North Atlantic having narrow and otherwise obstructed passages leading to its northern seas. Yet the high sea-level of the equatorial calm belt is always ready, whenever a favorable grade is formed by a monsoon or otherwise, to run off its surplus water obtained by winds and rain; and I have noticed, while cruising in these seas, that it happens at times during the northern winter months when the north-westerly gales drive the surface waters of the North-western Atlantic toward the tropical zone, and at the same time a strong north-east monsoon is prevailing along the southern coast of Brazil, the westerly currents setting past the Amazon River are reversed, and set to the south-east, while such conditions last.

For, when the summer solstice is in the south, and the north-east monsoon moves southward along the coast of Brazil, much equatorial water moves off in that direction; and during the same season the cooled Sahara has an outward flow of air toward the south, which moves more or less water from the coast of Guinea, which is easily accomplished, because the warm surface waters of that coast are inclined to join with the south equatorial stream. Consequently, the waters move from their high sea-level north of Cape Palmas, and so form the Guinea current.

The high sea-level of the equatorial calm belt of the Atlantic contains a large portion of the conserved heat of the tropical Atlantic, which at this age sends off a somewhat limited supply of warm water to the Gulf Stream, and also to the Brazil current. But, whenever the Cape Horn channel is closed or much obstructed, so causing a great low sea-level in the Southern Atlantic, the tropical waters heaped against Brazil, and the raised waters of the great calm region being one continuous high sea-level, would mostly be attracted to the vast low sea-level of the southern ocean. Hence it will be seen how large a portion of the conserved heat of the tropical Atlantic would be used to warm the high southern latitudes during a warm period in the southern hemisphere, and at the same time the head-waters of the Gulf Stream would obtain the same height as now. For we now see much of the force of the north-east trade winds lost, while maintaining so large a high sea-level to the windward of the West India Islands, which is probably capable of supplying a stream of double the capacity of the gulf current which passes through the Florida channel.

And it appears, while viewing the vast reservoirs of warm water apparently gathered by trade winds to subdue the cold of the high latitudes, that much of the energy of such winds is now lost to the world, while maintaining a vast and pent-up high sea-level which has a difficult outlet to the northern seas, and no strongly attractive low sea-level to move its waters into the oceans of the high southern latitudes. The wide waters which are banked up to the windward of the West India

Islands, and cause the wide tide-rips, set mostly to the westward into the Caribbean Sea through the passages south of Guadeloupe, while the northern portion of the raised waters set mostly toward the north, and thus form the eastern boundary of the Gulf Stream, and comprise the inner circle of the great current that encircles the Sargasso Sea.

I have been informed by an old Barbuda fisherman that "the weeds which float on the surface of the Sargasso Sea grow in large quantities on the bottom of the shoal waters to the north and eastward of that island and Antigua." Consequently, the currents of that region carry such weeds as become detached from their places of growth into the higher latitudes, where the westerly winds in the winter season drift them eastward south of Bermuda, until finally the central area of their gathering, where the most dense collection of weeds is found, is situated near the tropic of Cancer, and about 55° west longitude, as shown on map No. 2.

This position is also the centre of the great circular currents which encompass the Sargasso Sea. The comparatively few weeds which enter the Gulf Stream abreast Florida are currented to the northward of the Bermuda Islands, and from thence drifted by the westerly winds to the south-west of the Azores before entering the trade-wind belt. The weeds, on their long drift from their native shoals, hold their freshness, and continue to grow while floating on the sea for a considerable time, but at length lose their renovating properties, and in certain areas of the sea acquire an appearance of age and decay.

The Gulf Stream, and such other tropical waters as are attracted northward to the low sea-level abreast the North American coast, pass into the westerly wind-belt, and so gradually become drift currents, while being forced by the winds over to the European side of the ocean, as we have previously shown.

The vast movement of the North Atlantic waters encircling the great Sargasso Sea has often been pointed out by writers on the subject. But the central and most dense portion of the vast sea of weeds has always been placed on the charts several degrees of longitude east of its true position.

It is fifteen years since I wrote of the Gulf Stream and arctic currents as being attracted to a low sea-level caused by the westerly winds. But, as far as I know, writers on the Atlantic currents have had nothing to say of the great low sea-level caused by the westerly winds blowing the surface waters of the North Atlantic away from the eastern coast of North America, from Georgia to Newfoundland, and thus attracting the arctic and Gulf Stream waters in opposite directions, fifteen hundred miles along the North American coast. For, were it not for this low sea-level, the Gulf Stream would not be able to move so far northward as it now flows, but would spread out, were there no unevenness in the sea-level of the Atlantic, and become a drift current far south of its present northern limits. The United States government has caused surveys to be made of the Gulf Stream, and the interesting discoveries thus obtained have all been laid before the public. Still, such surveys cover but a portion of the whole round of the vast movement of the Gulf Stream water, and do not refer to the vast high sea-level of the calm belt as being one of its feeders, or to the wide disturbance of the surface waters of the tropical North Atlantic in their conflict with the trade winds, while being forced to the vast high sea-level of the Caribbean Sea and Gulf of Mexico, and so giving head to the Gulf Stream.

Thus from the foregoing explanations it will be seen that the ability of the prevailing winds to move the surface waters of the ocean away from the weather shores of continents over against the opposite leeward shores in the different wind-belts of the globe, and so cause both high and low sea-levels, is the main reason why there is an interchange of surface water between the tropical and colder zones sufficient to carry heat from the tropics to the cooler regions, and thus largely affect the temperature of the higher latitudes.

The unmistakable traces of cold periods having occurred in both hemispheres have given rise to an ingenious astronomical theory to account for their origin. According to this theory the ice periods in the two hemispheres were consecutive; and it is admitted by its supporters that, should it be shown that the frigid periods in the northern and southern hemispheres were concurrent, the astronomical doctrine would have to be abandoned.

It is impossible for a person who is acquainted with the great surface currents of the several oceans to conceive how a mild period could be maintained in the northern hemisphere with a frigid period existing in the southern hemisphere. A frigid period in the latter hemisphere necessitates a cold temperature for the superior oceans of the globe south of the equator. With this vast area of water reduced to a chilling temperature, it seems impossible for the inferior waters of the northern latitudes to maintain sufficient warmth to favor a mild period in the northern hemisphere, especially with both hemispheres receiving an equal annual amount of the sun's rays. The great Humboldt current, having its rise in the southern ocean west of Cape Horn, would during a southern frigid period greatly lower the temperature of the vast equatorial stream in the Pacific Ocean. Consequently, the Japanese stream, which branches off from the equatorial current into the North Pacific, would be cooled to such a degree that it would be unable to maintain the mild climate on the shores of the North Pacific which extensive lands now enjoy. Furthermore, during a cold period in the southern hemisphere the temperature of the Gulf Stream would also be greatly lowered by the great South-eastern Atlantic return current, which is caused by the south-east trade winds impelling the surface waters of that region into the equatorial latitudes, such waters being replenished from the common level of the southern ocean, and so mingling the cool waters of that sea with the equatorial waters of the Atlantic during a frigid period in the southern latitudes. And it may be said that during such times the frigid Antarctic Ocean would send its cold under-currents to cool the inferior northern oceans. Even to-day the northern and southern hemispheres, through the intermingling of the waters of the northern and southern oceans, largely maintain a like temperature in their temperate zones. Therefore, when we consider the certain traces of ice-sheets having formed on South Africa and Southern Australia, and to have overrun South America above the latitude of 40° south, thus strewing the oceans of the southern temperate zone with ice that are now largely free from it, it seems that the maintenance of warm oceans in the northern hemisphere during the time of a frigid period in the southern hemisphere would be impossible.

In order to make this statement more plain, I will again refer to the importance of the great Humboldt current for cooling the waters of the North Pacific during the perfection of a southern ice age. For during such times the ocean strewed with ice west of Cape Horn, where the Humboldt current takes its rise, would impart its coldness to the Humboldt stream, while it was floating icebergs toward the equator. The equatorial current of the Pacific being a continuation of the Humboldt stream, its waters would partake of its coldness. The Japanese current, being a large offshoot from the equatorial stream, would also possess a lower temperature than it obtains at this age. Yet at this date, with the southern ice-sheets confined to the antarctic lands, it does not possess heat sufficient to prevent glaciers from flowing down to the tide-water from mountains in Alaska.

Consequently, the Japanese stream could not maintain a mild climate on the North Pacific coasts while a cold period was being completed in the southern hemispheres. Therefore, under the conditions above set forth the support of a mild period in the northern hemisphere during the existence of a frigid period in the southern hemisphere could not be carried out.

From what has been explained, it will be seen that the growth of an ice period is necessarily slow, especially in its early stage, and also that the storage of ice is carried on in both hemispheres at the same time; but I will call further attention to the southern hemisphere, because it possesses greater resources than the northern for the production of an ice age.

The independent circulation of the southern ocean waters, as before shown, turns away the tropical currents, and thus largely prevents their warm waters from entering the high southern latitudes. Consequently, the heat from the sun's rays, and all other sources of heat included, are not sufficient to prevent ice from gathering on lands within the antarctic circle. This increasing storage of ice is only another name for the accumulation and spreading of cold, and so the increasing chillness goes on. The snow falls, and thus adds to the extension and thickness of the ice-sheets; and at the same time the spreading snow-fields reflect the heat received from the sun's rays into space, while the cold is retained and increased in the growing glaciers.

The spreading ice-sheets having covered the land are able to flow into the surrounding seas, where their outer edges become detached and form icebergs, which float out to sea, and so scatter over the adjoining oceans. Thus their coldness is mingled with and largely preserved by the sea, while the surface water, which is carried into the southern latitudes from the northern oceans by the prevailing winds, and also such surface waters as are attracted into the antarctic seas because of the difference of temperature of the antarctic waters and the more northern seas, are on gaining the frigid latitudes made cool, and returned to the more northern seas in cold under-currents, and so chilling the vast under-waters of the great oceans of the globe, and eventually their wide surface waters also; and so the coldness increases until the ice-sheets which at first formed on polar lands are enabled to spread slowly toward the equatorial regions so long as the independent circulation of the southern ocean is maintained.

But at length the depth of the great southern ocean is diminished because of the water evaporated from its surface, and precipitated in the shape of hail and snow over the vast continents and islands of the high northern latitudes, thus adding sufficient weight to the northern lands to attract the waters of the southern seas and still further lessen their depth. Thus during such times the Cape Horn channel is so reduced as to be obstructed by the heavy glaciers and icebergs of an ice age.

Consequently, a great change is wrought in the circulation of the southern seas. For, when the Cape Horn channel is closed, the westerly winds employ their strength to force the ocean's surface waters away from the glaciers which have filled the diminished channel. This potent action of the winds necessarily creates a great low sea-level on the Atlantic side of the obstructed strait, sufficient to attract the tropical waters heaped against Brazil by the trade winds, and the waters of the high sea-level of the equatorial calm belt, and also the equatorial waters which set along the east coast of Africa, well into the southern seas.

It will thus be seen that the conditions for the circulation of the tropical ocean waters have met with a great change.

But the temperature of the waters has been lowered by the coldness of a frigid period; and, consequently, their capability for conveying heat to the high latitudes has largely diminished. Therefore, their first inroads in the higher latitudes make small impression on the icy seas, so the early process for melting ice is exceedingly slow. But the icy southern ocean, deprived of its independent circulation, in the course of time yields to the warming invasion of the tropical waters, whose wide and increasing spread is eventually able to bring about a mild period, according to the natural methods which I have explained in the preceding pages.

And it may be said that a mild period succeeding a glacial age gained sufficient warmth to melt the ice-sheets from all lands excepting the highest mountains. For it is probable that there are lands situated in the antarctic circle sufficiently elevated even during late Tertiary times to have been above the snow-line. Therefore, the glaciers on such lands could not have melted away during mild periods succeeding an ice age. For, as has been explained, a portion of the waters of the southern seas had moved into the northern hemisphere. Consequently, the antarctic lands were raised higher above the sea-level than at this age. Hence the area of lofty land was increased above the snow-line. And, according to Dr. James Croll's estimate, the ice-sheet at the south pole is at this age several miles in thickness. Therefore, its upper surface is above the line of perpetual snow, and could not be melted away during the warm eras succeeding glacial periods.

CHAPTER III.

THE SPREAD OF GLACIERS
DURING COLD EPOCHS.

Spread of glaciers in tropical zone; Professor Agassiz on the origin of
Galapagos Islands; the bowlders of Hood's Island and rookery of
Albatross; alpine flora of Galapagos and tropical America; Mr. J.
Crawford on ancient glaciers in Nicaragua; Cuba and Republic of
Colombia during ice age; destruction of animal life during glacial
age; temperature of North Atlantic and Mediterranean Sea during
ice age; temperature of ocean during warm epochs; generative
age ascribed to warm eras; Professor Wright on pre-glacial man.

I HAVE before explained that the conditions are such that the cold periods of the
northern and southern hemispheres were concurrent. Through this cause, while
the glacial epoch was being perfected, the ice followed down the mountain ranges
of both hemispheres; and, while gathering on the lands of the temperate latitudes,
it also spread over a portion of the tropical zone. It is reported that traces of ancient glaciers are found in India, and also in Central America and in tropical South
America. In fact, the denudation caused by ancient glaciers on the elevated lands
of the tropics are too well defined to be attributed to any process of weathering,
while Alpine plants of the same species are found near the summits of mountains
in the tropics as well as in the high latitudes of both hemispheres.

This fact goes to show that a portion of the lowlands of the tropical zone have
experienced a temperature favorable for the growth of Alpine plants. And, judging from the tropical islands I have visited, situated in the cold currents which
flow down the eastern sides of the oceans from the high latitudes, I think they
show strong traces of having during some remote period been subject to the action
of glaciers. The island of St. Helena, situated in the southern tropical Atlantic, has
the appearance of having been heavily iced during a frigid age. Its steep ravines,
which deepen as they approach the sea, recall to the southern voyager the ice-
worn islands of the high latitudes. It seems improbable that these deep ravines
which penetrate the hard volcanic rock, on their short course to the sea, could have
been caused by their scanty brooklets.

The bowlders scattered over the island are not in harmony with the weathering process, while the obliteration of its craters seems to point to a more rapid
process of erosion than could be attributed to weathering.

Professor Agassiz, in his "General Sketch of the Expedition of the 'Albatross,'" states that the Galapagos Islands are of volcanic origin, and that their age does not reach beyond the earliest Tertiary period; and his report seems to favor the impression of their having undergone denudation sufficient to slough off large portions of the rims of the older craters, and also the eastern face of Wenman Island. On Hood's Island, at the time of my visit, its crater had entirely disappeared.

The highest portion of the island, which was the probable site of its ancient crater, showed no trace of its former existence; yet at the foot of this low mountain, on its southern side, I saw a large collection of loose bowlders, composed of hard volcanic rock, which were mostly free from soil and other débris, and easily moved from their places, while the spaces afforded by the loose piles of dark basaltic rocks afforded a secure retreat for numerous owls and lizards. Beyond the rocky piles to the southward a horizontal area of land was strewn with bowlders to the sea, which was some two miles distant from the higher land. The bowlders which covered the plain were somewhat smaller than those at the foot of the mountain, as none of the former were more than three or four feet in their longest measurement.

They seem to have been formed from thin strata of lava, which were broken in pieces from pressure, such as the action of ice could perform. In fact, the crowded and angular and somewhat worn blocks of lava presented a different appearance from stones thrown from the crater of a volcano, while no such bowlders are found among the recent volcanic eruptions on the islands.

The plain so thickly strewn with bowlders, and partly shaded by a tall growth of shrubs, fell off abruptly at the seaside, forming a steep cliff some two hundred feet in height.

The rocky floor at the foot of the cliff received such débris as fell from the sea-washed land; yet it contained few bowlders, they having been washed away by the waves soon after falling.

At one place a steep, dry ravine penetrated the land from the seashore, which was dangerous to cross on account of the loose stones resting on its sides. Two or three miles further west, on the level land bordering the sea, a large rookery of albatross were brooding their eggs and chicklings. The land on the south side of Albemarle, near the sea, consists of débris from the eroded high lands; and, judging from the crumbling cliffs by the sea, it seems that the land at one time extended further seaward.

Besides the excessive denudation which appears to have taken place on portions of these bowlder-strewn lands, we have other unmistakable testimony of their having formerly possessed a frigid temperature. The characteristic Alpine flora of these islands points to a time when they were exposed to a cold climate. Furthermore, rookeries of seal and albatross, which naturally belong to shores situated in cold latitudes, still exist on these equatorial islands; and, when we consider the favorable position of the Galapagos for the reception of cold during a frigid period, we can well account for the lingering signs which point to their former cold climate.

During the perfection of an ice period the western shore of South America was covered with an ice-sheet from the summits of its mountain range to the sea, extending northward as far as the latitude of 38° south.

This vast ice-sheet, situated in a region of great snow-fall, was constantly sending icebergs into the sea, where they were borne northward by the cold Humboldt current directly toward the Galapagos Islands; while, on the other hand, in the northern latitudes, in regions of great snow-fall, such as Alaska and British America, numerous icebergs were launched into the ocean, to be currented southward to the Galapagos seas. Thus during the frigid epoch the equatorial waters surrounding the Galapagos group was one of the greatest gathering places for floating ice to be found on the globe.

And here the frigidity stored up in the glaciers of the higher latitudes was set free, thus chilling the waters as well as the atmosphere of that region. The Alpine flora of the American coast mountains was probably carried by floating ice to the Galapagos, while its rookeries of albatross and seal date back to a cold period. And it seems that these cold-weather animals, with the assistance of the cool Humboldt current, may be able to preserve their rookeries at the equator until the advent of another ice period. In connection with the evidences of a cold climate having possessed the Galapagos, there are ample traces of ice-sheets having flowed over a large portion of the high lands of tropical America, and in some places the ice may have flowed down to the sea, especially where the large rivers now empty; and it is said that masses of clay, mixed with sub-angular stones, have been found in Brazil, which goes to prove the glaciation of portions of that tropical land during a remote age. Professor Louis J. R. Agassiz, during his research in the Amazon valley, found bowlders resting near the summits of the low hills of that region, which he attributed to the action of ice. The spread of glaciers on southern continents and islands is shown on map No. 1.

In *Science*, Nov. 17, 1893, Mr. J. Crawford published a summary of his discoveries in Nicaragua, during ten months of nearly continuous exploration since August, 1892.

The author of this report says: "The numerous eroded mountain ridges and lateral terminal moraines of that tropical region give unquestionable evidences of the former existence of a glacial epoch, which covered an area of several thousand square miles in Nicaragua with glacial ice. The ice-sheet covered a large part of the existing narrow divide of land (containing about 48,000 square miles) between the Pacific and Caribbean Sea." And it is likely that other large areas of tropical America were glaciated at the same time, especially in regions of great precipitation.

The island of Cuba, during a portion of the ice age, probably supported heavy glaciers, and obtained an average temperature as low as South-western New Zealand at this age. According to the description given by J. W. Spencer, of the Cuban land, great valleys have been excavated, the lower portion of which are now fiords, reaching in one case at least to seven thousand feet in depth before gaining the sea beyond. Thus, while keeping in view the glacial condition of Central America during the frigid period, it seems that the great Cuban excavations were partly the work of glaciers of the same cold epoch.[1] Judging from such reliable statements, it is probable that the climate of tropical America during the frigid age was somewhat colder than obtained in the tropical regions of the eastern continent, owing to the wide connection of the Atlantic with the Arctic Ocean as well as with the antarctic seas, and because of its shores possessing a larger area of glaciated lands in proportion to its size than the Pacific and Indian Oceans, and also owing to the tropical Atlantic containing so small a portion of the world's waters which lie within the torrid zone, and its equatorial current being separated by continental lands from the great equatorial stream of the Pacific and Indian Oceans.

Therefore, the tropical Atlantic waters must have been reduced to a lower temperature during a frigid age than the tropical waters of the Indian Ocean or the western part of the tropical Pacific, as a large portion of the great equatorial current of the latter oceans, during its western movement, was exposed to the rays of a tropical sun for a much longer time, after being replenished by the cold waters of the high latitudes, than the tropical currents of the Atlantic; and it is probable that, on account of tropical America possessing a colder climate than the tropical lands of the eastern continent during the frigid epoch, the cold of the western continent was more destructive to its fauna and flora than was the case in the tropical regions of the eastern continent. Professor Wright, in his valuable work on "The Ice Age of North America," gives a good description of the "flight of plants and animals during the glacial epoch," and also of the extermination of many superior species because of the frigid climate.

The high lands of tropical Africa, above the altitude of three thousand feet, and situated in places of great precipitation, were probably covered with snow and ice during the glacial age. Travellers have reported that islands composed partly of granite bowlders are found in the lakes at the head-waters of the Nile. But the glaciers that invaded the tropical latitudes were of short duration compared with the ice-sheets that burdened the lands of the temperate zones. Besides, such tropical ice as flowed to the low lands was so near a melting condition that it made small impression on the rocks; but on steep mountain slopes, where the movement of the ice was comparatively rapid, it possessed considerable eroding power. The climate of the tropical zone on both continents during the perfection of an ice period was so cold that such animals as could not endure a low temperature retreated into the warmest regions of the equatorial latitudes, while many species who failed to reach such places perished. And especially was this the case with the pre-glacial fauna of the western continent. Mr. W. B. M. Davidson, in his treatise on Florida phosphates, says: "The great mammal hordes of the glacial epoch were driven into Florida in their flight southward for life and warmth, and there perished because of the deadly cold which ever moved southward. The Florida

---

[1] The meeting of the British Association for the Advancement of Science, September, 1895, was reported in *Science* of October 18, where mention is made of an interesting paper by Mr. R. B. White, on "The Glacial Age of Tropical America," in which he described a number of apparently glacial deposits in the Republic of Colombia, almost under the equator. He spoke of moraines forming veritable mountains, immense thicknesses of bowlder clay, breccias, cement beds, sand, gravels, and clays, beds of loess, valleys scooped, grooved, and terraced, monstrous erratics, and traces of great avalanches.

waters grew so icy cold, fishes, reptiles, and mammoth animals died, and added their frames and teeth to the valley of bones now found in that southern region."

Such species of the tropical fauna of the ocean as survived the ice age could have existed only in torrid seas with small connection with the cold oceans during the frigid epochs. For, with the diminished oceans of a cold period, it seems that the conditions were favorable for the maintenance of such seas in the region of the East India Islands.

Such parts of Southern Europe and Northern Africa as bordered on the Mediterranean Sea probably possessed a milder climate during the ice age than regions in the same latitudes on the Atlantic coast, for the reason that the North Atlantic was proportionally a greater receptacle for icebergs which were launched into it from the numerous glaciers of North-eastern America, Greenland, Iceland, and North-western Europe than the great inland sea obtained from its less frigid shores. And it may have happened that during such times the tropical waters of the Indian Ocean had some connection with the Mediterranean through the Red Sea and Suez, and so during portions of the year the waters of the tropical Indian Ocean were forced by the periodical winds into the inland sea. It is the opinion of several writers that man, along with other species of animal life, existed previous to the glacial period; for, since the seas and lands of the globe were chilled, the conditions seem to have been less favorable for the spontaneous generation of animate bodies than during the previous warm ages. Therefore, it appears that the generative ages should be ascribed to the long genial eras prior to the glacial epochs. For it is probable that the lower parts of the ocean, which now possess a low temperature even in the tropical latitudes, were, during the warm eras, wholly composed of warm water, because the surface waters of the antarctic seas of that age, which supply the great under-currents of the ocean, would possess a high temperature; and it is probable that the temperature of a large portion of the seas of the torrid zone was for a long time maintained at blood heat. For it should be considered that the waters which moved from the torrid seas, after making their journey through the warm regions of the high latitudes, would on their return to the tropics retain a large portion of the heat they acquired in the torrid zone before making their journey to the mild polar regions.

And, when we reflect how the heat of the sun's rays was conserved by the ocean waters, and that their circulation during such times was almost wholly performed by the winds, as the difference of temperature between the polar latitudes and the equator was small, it appears that during the eras previous to the glacial age the oceans must have obtained a higher temperature than possessed by the warmest seas of to-day.

According to the discoveries of Professor Wright and others, ancient stone implements have been found beneath the glacial drift, as well as the bones of animals whose descendants are now living, which goes to prove that man, with other species of fauna which now inhabit the earth, existed anterior to the glacial epoch.

And on consideration it seems unreasonable to suppose that any of the superior species of animals could have been brought into existence since the waters and lands of the earth were chilled by the cold of a glacial age. And it appears that many species of animals which are known to have survived the cold periods were indebted for such survivals to the slow process through which a frigid period is brought about, thus affording time for evolutionary inurement to the slow increase of cold which at length perfects a glacial epoch.

The inurement to cold acquired by animals during the glacial age is still an attribute possessed by many species of fauna to-day. For, when a warm climate took possession of the tropical zone, it was deserted by a large portion of the animals that found refuge there during the glacial age.

Thus, while the seas and shores of the cooler latitudes swarm with animate bodies, the torrid latitudes seem comparatively lonely to the voyagers on the tropical oceans.

CHAPTER IV.

THE GLACIERS OF THE TEMPERATE ZONES.

Professor Hitchcock on the early history of North America; glacial deposits of Nantucket and Martha's Vineyard; Professor James Geikie on the glacial deposits of Northern Italy; California coast ranges the work of Sierra glaciers; ancient glaciers on the Pacific slope north of California; Professor Geikie's views on the ancient glaciers in the Salt Lake region; Colorado Cañon; the conglomerate deposits in the Appalachian district; remarks on the glacial boundaries in United States during ice age; sands of Florida; ancient ice-sheets of the plains west of the Mississippi River; the driftless region of Wisconsin; tropical waters of North Atlantic chilled during ice age; the drifted snow of British America and Siberia during ice age.

HAVING asserted that during the culmination of a frigid period the ice-sheets spread over a portion of the lands of the tropical zone, I will give my views, with those of several writers, on the spread of ice-sheets within the now temperate latitudes; and meanwhile I will repeat a portion of my former essays on the subject. Professor Hitchcock, in his lectures on the early history of North America, says that "the history opens with igneous agency in the ascendant, aqueous and organic forces become conspicuous later on, and ice has put on the finishing touches to the terrestrial contours." But there appear to be various opinions held by geologists respecting the changes brought about on the earth's surface during the glacial period. Some think that glaciers have never been an important geological agent, while others assert that during the glacial epoch heavy ice-sheets covered the elevated portions of Western North America as far south as the thirty-sixth parallel of latitude, and Eastern North America was overspread with ice-sheets, which attained a depth of five or six thousand feet, and were able to move their débris over wide lands of little declivity toward the sea, their immense deposits forming the lands of Cape Cod, and also the islands of Nantucket and Martha's Vineyard.

But it is now said that this implied magnitude of the glacial deposits on the lands skirting the New England coast is without foundation, since the larger bulk of these islands consists of upturned Cretaceous and Tertiary strata, which are only thinly covered with glacial débris, such as bowlders, gravel, clay, and sand, from the eroded shores of the mainland of New England. But it appears that the dislocated and folded cretaceous strata which underlie the glacial drift of Nantucket and Martha's Vineyard were during an early period deposited on the bottom of a shallow sea, which then covered the Vineyard Sound, Buzzard's Bay, and their surrounding lowlands. Thus the ice-sheets of the frigid age which moved over New England displaced the yielding stratified deposits of the shallow sea, and forced them southward in a disturbed condition to the position which they now occupy.

Still, it is apparent that only a small portion of the glacial drift is found on these islands, which, according to appearances, must have been eroded and moved southward from the rocky lands of New England during the ice age; but there is sufficient to show that large quantities of such débris were carried over the islands into the Atlantic. And, judging from the eroded rocky New England lands, there must have been sufficient glacial drift moved over Nantucket and Martha's Vineyard into the ocean beyond to far exceed in bulk the deranged Tertiary and Cretaceous deposits which now form so large a portion of the islands.

For, when we look over lands bearing traces of the ice age, where the glaciers did not move their drift into the sea, so the terminal moraines of such glaciers can be better estimated, we can realize the great work that has been performed by the ice-sheet that overran New England during a frigid age.

Professor James Geikie states, in his discussion on the glacial deposits of Northern Italy, that the deposits from Alpine glaciers of a frigid period "rise out of the plains of Piedmont as steep hills to a height of fifteen hundred feet, and in one place to nearly two thousand feet. Measured along its outer circumference, this great morainic mass is found to have a frontage of fifty miles, while the plain which it encloses extends some fifteen miles from Andrate southward." And it is reported that there are found on the southern flank of the Jura numerous scattered bowlders, all of which have been carried from the Alps across the intervening plains, and left where they now rest. Many contain thousands of cubic feet, and not a few are quite as large as cottages.

Such blocks are found on the Jura, at a height of no less than two thousand feet above the Lake of Neuchâtel. The Jura Mountains being formed of limestone, it is easy to distinguish the débris deposited by Alpine glaciers; and, from what I can learn of extensive glacial work, it appears that intervening plains, lakes, and sounds are so often found separating the source of ancient glaciers from their deposits that their existence becomes almost necessary to represent the general outlines of disturbance performed during an ice period. In consideration of such facts and the foregoing statements of reliable observers, I am prompted to offer my views on glacial work performed on a portion of the Pacific shores of North America, which seems to me to be much more extensive than hitherto supposed.

Professor Whitney describes the coast mountains of California as being made up of great disturbances, which have been brought about within geologically recent times; and this statement I found to be so obvious in my travels over that region that it appears to me that the coast ranges originated in a different manner from the older Sierras. The western sides of the latter mountains everywhere show the great eroding power of ancient glaciers; and, when I considered their favorable position for the accumulation of snow during a glacial period, I was led to seek for the glacial deposits adequate to represent the great gathering of ice which an age of frigid temperature would produce.

But it seemed to me that such deposits could not be found in the foot-hills of the Sierras, which contain the moraine of inferior ice-sheets that terminated at the base of the mountains.

Under these conditions I came to the conclusion that during the earlier ice period the immense glaciers which must have formed on the western slopes of the Sierra range moved their gigantic accumulation of débris so far seaward as to form the range of hills now existing next the coast line, and perhaps the islands abreast the Santa Barbara coast, the Contra Costa, or eastern range, being formed during a subsequent ice period, in the same manner as the hills next the coast line.

Still, it may be that neither of the coast ranges was the work of a single cold epoch; but the western range must necessarily have been the earliest deposit. Although the coast ranges differ from the Sierras in their make up, yet it does not disagree with the glacial origin of the former inferior mountains, from the fact that the ice-sheets, while moving their bulk westward, displaced the deposits of such bays, lakes, rivers, and marshes as lay abreast of the Sierra slopes. The advancing ice-sheets, thousands of feet in depth, moving from a lofty and steep incline, pressed and ploughed below the somewhat superficial cretaceous and alluvial strata which lay in their course. The disturbed strata, while forced along in confused heaps in front of the ice, were amassed in ridges sufficient to form the hills of the coast ranges. The bowlders found imbedded in several of the coast hills must have been moved by the ice from the Sierras on account of the coast ranges not having a rocky core of sufficient firmness to give shape to such bowlders. Moreover, the temperature of the Pacific waters would not be favorable for glaciers to form on the coast ranges, with the ice-sheets of the Sierras terminating at the foot-hills.

The Sacramento and San Joaquin Valleys are now covered by recent river deposits. Therefore, the glacial drift which should be traced from the Sierras to the coast ranges is concealed.

Yet the abraded appearance of exposed solid rocks at the base of the foot-hills, and also the scattered bowlders which gradually disappear beneath the diluvial deposits of the plains, indicate that the Sierra ice-sheets could not have ended at the foot-hills, but must have moved further westward, while gathering immense accumulations in their front, sufficient to form the coast hills, the débris thus amassed being able to arrest the further movement of the ice seaward.

The coast ranges in several places have been subject to igneous action, which may have been brought about through heat generated from pressure exerted on the interior masses after the ice had melted away, the heat thus produced being sufficient to cause outbursts of lava, where the nature of the material favored combustion. The low plains, lakes, and bays which separate the Sierras from the coast hills are in a position similar to the shallow sounds which separate Nantucket,

Martha's Vineyard, and Long Island from the inferior slopes of the mountains of New England. Therefore, while agreeing with glacialists, who believe that great geological changes have been wrought by ice-sheets in Italy and New England, it appears to me that the ancient glaciers of the Sierra Nevada have accomplished more extensive work, owing to the Sierras being situated in a more favorable position to receive the humidity of the ocean.

Hence, with a low temperature, vast quantities of snow must have collected on their lofty sides; and at the same time their great height and declivity would cause the ice to move down their steeps with greater force than the glaciers which passed over New England. Writers who have given the subject considerable study think that the deep valleys of the Sierra Nevada were produced by disruptive rather than erosive agencies. This conclusion has been formed from the lack of large accumulations of débris about their lower extremities, which would not be the case if such valleys were the result of glacial erosion. But, should the coast ranges be attributed to glacial action, as has been stated, we can well account for the débris that should accumulate from the erosion of the deep valleys.

The only thing that could prevent the ice from gathering on the Sierra Nevada range during an ice period in greater masses than on any mountains in the northern hemisphere would be the lack of cold; for, with a low temperature, the fall of snow would be enormous. This is shown by the great snow-fall during the short mild winters of to-day. Therefore, with ice-sheets covering a large portion of the lands of the high northern latitudes, and with the Japanese current which tempers the north Pacific waters made cold in the manner described in the foregoing pages, and while the sea along the north-west coast of America was strewn with icebergs launched from Alaska and British Columbia, it seems that California must also have obtained a frigid climate during the ice age. Therefore, on account of its exposure to the ocean winds, and the consequent heavy snow-fall, the accumulation of ice on its lands must have been immense. For, when it is considered that the glaciers of North America extended southward even into the torrid zone sufficient to cover a large portion of Central America, it is unreasonable to suppose that any portion of California could escape being covered by heavy ice-sheets during the glacial epoch. The comparatively scant fall of rain and snow over Greenland is known to form ice-sheets hundreds of feet in thickness.

Therefore, what must have been the depth of ice over the high lands of the Pacific coast north of California at the culmination of a frigid period? The descriptions given by Dr. Dawson and others, of glacial phenomena along that coast, favor the impression that an immense ice-sheet at one time deeply covered the whole region from the top of the mountain range to the ocean.

Thus all the deep channels were filled and all the islands deeply overrun with ice, while the immense bergs launched from the shore and carried by the winds and currents southward were probably not melted until they reached the tropical latitudes. Thus, when the whole circulation of the Pacific waters are taken into account, it will be seen that their temperature during the ice age must have been considerably lowered. The movement of ice-sheets on the Pacific slope was probably local in character, and not connected with the movement of ice on the eastern sides of the mountains.

From what I have seen of the vast territory lying between the Sierra Nevada and the Rocky Mountains it appears that it obtained much heavier ice-fields than generally supposed. Professor Geikie in his lectures says of this region that during the glacial age, "in the Second Colorado Canyon, the sides were completely glaciated from bottom to top. These walls are from 800 to 1,000 feet high, and at the thickest point the glacier was 1,700 feet thick"; and he says that "the country around Salt Lake was covered with ice, for the rocks about there show the action of ice, and that the bones of the musk-ox are found there." This vast area of ancient ice, although subject to little movement in its interior basin, still, in whatever movement it may have had, must have found its main outlet through the Grand Canyon of the Colorado.

For in no other way can we account for the erosive forces necessary to excavate that immense chasm. Not even the mighty torrent that carried off the waters of the melting ice-sheets that covered the interior portion of the continent could accomplish work of such magnitude.

According to Professor Geikie's observations the Second Colorado Canyon was filled with glaciers during the ice age. Therefore, it seems that these glaciers must have flowed down into the Grand Canyon, and there united with glaciers flowing from more northern regions.

An account of a collecting expedition to Lower California by G. Eison, in 1895, describes ancient moraines at the extremity of the peninsula as being prominent, large, and steep. This region lies under the tropic of Cancer, and 8° south of the mouth of the Colorado River where it empties into the Gulf of California. Hence it appears that the temperature of that portion of North America during the ice age was favorable for the great glacier of the Colorado Canyon to have flowed into the Gulf of California.

The wide, shallow basins of Utah and Nevada were filled with the water from the melting ice-sheet on the breaking up of the ice period, and the lakes so caused remained for a considerable time after the disappearance of the ice. But, owing to the great evaporation and light rain-fall of that region, the lakes gradually shrank away, the filling and emptying of the lake basins being governed by the cold and mild epochs.

The conglomerate deposits in the Appalachian district of North America are known as occurring on a large scale. Professor Shaler is inclined to attribute them to glacial action, because he knows of no other force that could bring together such masses of pebbles from a wide-spread surface. In Eastern Kentucky and East Tennessee these deposits are found to be several hundred feet in thickness. Such accumulations of apparent glacial origin are to be found from New Brunswick to Alabama.

Hence it seems that the ice during a frigid period followed down the Alleghany range even so far south as Georgia and Alabama; and for a time, when the ice attained its greatest spread, it flowed over the central portion of the Gulf States. For how else can we account for the clay mixed with gravel and pebbles and stony fragments being spread broadcast over that region?

I know that such statements do not agree with the views of glacialists who have written on the subject, and have drawn the glacial boundary from seven to ten degrees further north, where a line of bowlders with other glacial débris is plainly traced. Still, it appears to me that a line of bowlders deposited by an ice-sheet spreading over a continent and across many degrees of latitude cannot be compared to the moraines of inferior mountain glaciers of the temperate latitudes of the present age.

An ice-sheet moving from a high latitude to a lower would, while in the colder latitude, freeze firmly to the rocky ledges, and hold them so strong in its frigid grasp as to break off the weaker portions of the rocks, and drag them toward a milder region, as far as the freezing grip of the ice-sheet would permit; but, on gaining lower and milder latitudes, the holding and dragging power of the ice would be lost on account of the increased warmth of the earth over which the glacier must pass, and also because of the ice-sheet having lost a portion of the low temperature acquired in the higher latitudes. Therefore, on such lines the bowlders would be released, while the ice-sheet would still move on, although largely deprived of its eroding power.

This is the probable reason why a line of glacial débris, largely composed of bowlders, is found to extend across the Middle and Western States, and so generally supposed to be the glacial boundary of a frigid period. But there is no reason to suppose that an ice-sheet, although deprived of its eroding power, was arrested in its southern movement on the line of its stony débris, because there could be no sudden change of temperature in a particular latitude on the eastern lands of North America to cause an abrupt ending of the ice-sheets. And there appears to be nothing to hinder the ice from gathering and flowing over lands warm enough to loosen its implements of erosion; for there is much to show that the ice-sheets flowed much further southward, even into the middle portion of the Gulf States, and there spread the clay mixed with gravel and pebbles, with now and then a bowlder, over the land. The scattered bowlders, found in numerous instances many miles south of the bowlder line, were so deeply imbedded in the ice-sheet that they could not be dropped on the usual releasing ground. The ice-sheet, when deprived of its rocky, eroding implements, would, while flowing over the land, leave few or no imprints on the rocks; but it would probably move and spread a large amount of clay, gravel, pebbles, and sand over its wide course, especially if the ice moved from a region abounding with such material.

Should we place the glacial boundary on the line of the rocky débris, how could we account for the glaciated stones found on the hills and plains situated far southward of the bowlder-strewn regions of the Middle and Western States? The clay mixed with gravel and sand, and spread so broadcast over a large portion of Georgia and even into Northern Florida, makes it appear that the ice of a cold period must have covered that southern region.

Moreover, it seems to have been through the great abrasion which only ice-sheets could perform that the sands of the Florida peninsula were produced; for on examination they seem to have resulted from the abrasion and weathering of crystalline rocks.

The worn remnants of such rocks are now found in the southern Appalachian range. In fact, the hills and mountains of that region at the present time are supposed to be a small remnant of the ancient highlands. Thus, on consideration, it appears that the sands caused by the action of glaciers were, on the disappearance of ice-sheets, blown by the strong north-west winds toward the Florida peninsula as fast as the receding waters of the ocean which flowed the lowlands on the breaking up of the ice age would permit; and in this way the sand was spread over the lowland region, which was largely composed of coral sea shells and other marine matter. And it seems that the sand must have been blown over large areas in Florida soon after the ending of the frigid period, because the sand, in order to be moved by the winds, must have spread over a country nearly destitute of vegetation; and such would be the condition of that region during times which succeeded the ice period and the subsequent brief flowage of the lowlands on the ending of the frigid age, which would not be the case if such sands resulted entirely from water erosion and weathering, because with such a state of things the country would be covered with forests and grasses, which would prevent the sand from being moved by the winds to any great extent.

This goes to show that the region of the Gulf States was so much affected by the cold of the glacial period, together with the submergence of the lowlands at its close, its flora and also its animals were exterminated; for how else can we account for the abundant fossil remains of animals now found buried in the Florida sands? It appears also that, when Florida was being covered with drifting sands, many of the lake basins now formed did not exist, as the wind-blown sand could not have crossed a continuous chain of lakes like the St. John's River; and it is an easy matter to-day to trace the beds of the ancient lakes that prevented the sands from drifting over certain lands now nearly destitute of it. And it is probable that the sea flowed the lowest lands during the period when the winds were drifting the greater portion of the sands over the peninsula. Therefore, regions which embrace the Everglades and portions of the Indian River territory are quite free from heavy sand deposits, and so also are the extensive flat woods of the peninsula.

Since the sands blew over the ancient desert of Florida, many lake basins have been formed because of the sinking of the ground. This sinking of the ground is a common occurrence in limestone regions, where a great amount of material is moved in solution, leaving caverns whose roofs often fall in. The great amount of sand blown upon Florida caused the marine strata to give way in the weaker places under its burden. The sinks thus formed, probably of frequent occurrence at one time, have now nearly ceased. Still, there are depressions to be seen to-day where the tops of large pine-trees, which grew on dry, sandy land, are barely above the surface of the water which partly fills the basins so recently formed. Yet I would not assert that all of the depressions where Florida lakes exist were caused by the sinking of the ground; for the winds may have caused shallow basins in the sand, where the decayed vegetation has formed mud sufficient to hold the water which now partly fills such basins.

The mobility of Florida sands can be seen to good advantage when exposed to a strong, dry north-west wind, where the ground happens to be destitute of vegetation. An observer can then realize what the result would be, should the whole land be deprived of vegetation and laid bare to the action of the winds.

Under such conditions, not only would the winds be much stronger than now, but the air near the ground would be filled with sand, moving like drifting snow in a Dakota blizzard. And, furthermore, it is probable that the rainfall was very light while Florida was void of vegetation; and, even if shallow basins were formed, there would be a lack of rain to supply them with water.

The wide plains west of the Mississippi River, extending southward into Texas, during the frigid period must have been covered with a sheet of ice and snow. And it is probable that it was not wholly a product of more northern latitudes, but was mostly produced by the snow which fell on the plains during the long winters of that period, which could not be melted away during the cold summers of an ice age, when it is considered that an ice-sheet, with a temperature sufficiently low as to carry glacial drift, covered the lands of Missouri as far as latitude 38° south; and it may have been through the pressure from an ice-sheet in its south-eastern movement that we are to account for the numerous ore-bearing faulting fissures traversing the limestone strata.

The ice-sheet was also the probable cause of the erosion of the horizontal bedded stones, yet it appears that the ice did not greatly change the contour of the ground; for it is well known that glaciers do move over lands that are not frozen to the ice without causing much disturbance, especially where the gradient is small, and this was the probable condition of the Western plains during the ice age. Thus it seems that whatever disturbance this region has undergone could be partly attributed to ice-sheets without the presence of bowlder drift, because the temperature and texture of the ground in the limestone region were unfavorable for such accumulations; yet it may be owing to the action of ice that minerals once diffused are now found collected in fissures. The deep valleys through which the large rivers now pass on their way toward the sea were once filled with glaciers which flowed into them from their tributaries. Thus the deep trenches of the plains are largely the work of glaciers. It is generally supposed that the driftless region of Wisconsin was free from ice during the frigid period. But it seems impossible for this region to have escaped being covered by ice and snow, with the great lakes filled with glaciers, and the regions on all sides of the driftless area covered with ice.

The reason why this territory escaped the drift from the north was on account of the hindrance which the drift-bearing ice-sheet encountered in the deep basin of Lake Superior. In this great depression the ice-sheet from the north was relieved of bowlders and other glacial drift, as well as obstructed in its southern movement.

Therefore, the snow and ice which gathered on the driftless region had little movement in any direction, while the temperature and consistency of the ground under the ice were not favorable for the production of bowlder drift; and, when we consider that the Mississippi valley was deprived of great sources of warmth during the culmination of a glacial period, we are forced to the conclusion that its wide lands were also covered with snow and ice.

The tropical waters of the North Atlantic were so much chilled by the floating icebergs of North-eastern America, Greenland, Iceland, and Northern Europe that the Caribbean Sea, its warmest reservoir, was reduced to a temperature so low that the easterly winds which blew over its waters were unable to prevent ice-sheets from gathering on Eastern Nicaragua.

Therefore, during such frigid times it appears that, with the waters of the Caribbean Sea and Gulf of Mexico reduced to a low temperature, it was impossible for the great Mississippi valley to escape glaciation, while being surrounded by cold seas and glaciated lands which extended even into the tropical latitudes. The broad, level lands of British America and Siberia during the ice age must have been thickly covered by the snow which fell on the deeply frozen plains, besides the large amount of snow that the cold westerly winds must have drifted over their icy surface from lands of greater snow-fall on their western borders. This snow during such freezing times could not be melted away.

The great ice-sheets thus formed over wide, level lands could have but little motion in any direction, certainly not sufficient to cause glacial drift of much magnitude; yet the ice-sheet, at one stage of its existence, probably served to widen and deepen the channels of the great rivers which empty into the Arctic Ocean from these vast regions, and the glacial débris from such erosion was deposited in the arctic seas.

CHAPTER V.

REMARKS ON THEORIES ADVANCED FOR EXPLAINING ICE PERIODS.

Professor Geikie on supposed causes of the glacial period; change in the relative level of the land and sea during glacial and post-glacial times; submergence of northern lands at close of ice age; the main cause of the movement of water from the northern seas at the close of glacial age; why the earth-movement hypothesis should be rejected; glaciers of Europe and Alaska; North Pacific currents; why the Pacific waters are growing cool; the lowering temperature of the northern seas; the increase of cold in Europe and Asia; falling temperature of the Andean region; General Drayson's astronomical discoveries for explaining the cause of ice periods; why the Gulf Stream was always confined to the North Atlantic; the improbability of the Indian Ocean currents entering the arctic seas; why the increase of glaciers must continue while the Cape Horn channel maintains its present capacity; comments on the coming ice age; tropical zone the abode of man during ice age; preservation of the tropical ocean fauna through the glacial period.

On Nov. 12, 1891, Professor Geikie made his presidential address before the Edinburgh Geological Society, the subject being "Supposed Causes of the Glacial Period."

Many of his views advanced in this lecture were so much in accordance with my own that I am induced to repeat them. He said that the glacial period was a general phenomenon due to some widely acting cause, and that where we now have the greatest rain-fall the greatest snow-fall took place, and that the Pleistocene period was characterized by great oscillations of climate, extremely cold and very genial conditions alternating. He also said that in glacial and post-glacial times changes in the relative level of the land and sea had taken place, and any suggested explanation which did not fully account for these various climatic and geographical conditions could not be satisfactory. And, while examining the earth-movement hypothesis, he pointed out that in the first place there was not the least evidence of great continental elevations and depressions in the northern hemisphere, such as the hypothesis postulated. Next he showed that, even if the disserrated earth-movements were admitted, they would not account for the phenomena.

Such changes, no doubt, would profoundly affect the maritime regions of North America and Europe; but they would not bring about the conditions that obtained at the climax of the ice age.

Another objection to the earth-movement hypothesis was this: it did not account for interglacial conditions. The advocates of that hypothesis imagined that these conditions would supervene when the highly elevated northern regions were depressed to their present level. But these were the conditions that obtained at the present time; and yet in spite of them the climate was neither so equable nor so genial as that which obtained in interglacial times and during the mild stage of the necessary post-glacial period.

Therefore, he said that the earth-movement hypothesis should be rejected, not only because it was highly improbable that such wonderfully rhythmic elevations and depressions of northern lands could have taken place, but chiefly because it did not explain the conditions of the glacial periods and interglacial times.

Still, Professor Geikie says that in glacial and in post-glacial times changes in the relative level of the land and sea had taken place; and it is reasonable to suppose that such changes were obtained in the high latitudes of both hemispheres during the breaking up of the last ice age.

We have previously pointed out that much of the ice of the glacial period in the southern hemisphere was melted away, and its waters warmed sufficiently to assist the Gulf Stream and Japanese current to bring about a mild period in the northern hemisphere; for without such assistance they would be unable to disperse the vast ice-sheets of the northern latitudes.

Still, the attraction of the southern ocean waters into the northern seas must have commenced as soon as the growing ice-sheets of the large continents and islands of the high northern latitudes surpassed the growth and weight of the glaciers on the smaller lands of the southern hemisphere.

Hence the attraction of the ocean waters northward overcomes the force of the prevailing winds from moving an undue portion of the ocean's surface waters southward. Consequently, the movement of water from the southern seas into the northern latitudes continued so long as the vast northern ice-sheets increased in weight greater than the glaciers of the southern hemisphere. Therefore, at the perfection of a frigid age straits and channels situated so far southward as the Magellan and Cape Horn channels were much diminished in width and depth or entirely deprived of their waters. Through this cause such reduced channels were readily filled with glaciers in a region of great snow-fall. The depth of water on the submerged northern lands at the close of the glacial period is not known.

According to Professor Dawson, in the township of Montague in Ontario the skeleton of a whale was found in post-glacial deposits 440 feet above tide-water, and marine shells are known to occur on Montreal mountain at an elevation of 520 feet above the ocean; and it is said that there are traces of submergence of over one thousand feet in the higher latitudes, including the islands of Great Britain.

According to the researches of Dr. J. W. Spencer, one great sheet of water covered most of the great lake region about the close of the ice age; and the lower strands of these inland seas are known to be connected with old marine shore lines. The probable reason why so few sea-shells collected on the glacial drift during such times was because of so much marine life having been exterminated in the high northern latitudes during the frigid age. Therefore, the sea, in the short period of northern submergence, left but few traces on the glacial drift it once flowed.

Thus it will be seen that, if the ocean waters were attracted northward through the preponderance of northern ice-sheets, they not only assisted in melting the northern ice, but also served to greatly reduce the waters in the Cape Horn channel, and so largely prevented the independent circulation of the southern ocean, thus furthering a mild climate in the southern hemisphere until the prevailing winds, after the northern ice-sheets were melted, were able to move more of the ocean waters southward than they could move northward, owing to the ocean currents setting southward being less obstructed than the lesser currents setting northward. This tendency of the ocean waters to move southward I have before explained in the preceding pages.

But I will say in addition that, on further consideration, it seems that one of the main causes of the waters of the augmented northern oceans moving southward so soon after the melting of the ice from the northern lands was on account of so much water being attracted southward to the great low sea-level east of Cape Horn. This vast low sea-level remained a great area of attraction for the northern seas until so much northern water was moved into the southern ocean as to reduce the seas of the northern hemisphere and augment the southern ocean sufficiently to enlarge the Cape Horn channel, thus causing the extinction of the vast low sea-level that furnished such great attraction for the waters of the more northern latitudes.

If the earth-movement hypothesis, so wholly rejected by Professor Geikie, fails to explain the cause or causes of a northern ice age, it seems to be still more inadequate for explaining the occurrence of ice periods extending over both hemispheres. For it is not probable that portions of continents and large islands rose above the snow-line in both temperate zones during the same period of time, and then again obtained their present level with the occurrence of a mild era.

Those who maintain that the continents of North America and Europe rose to great elevations during the ice age, in order to prove their assertions, point to the fiords which indent the eastern and western coasts of North America, and also to the fiords of Norway, as having been eroded by streams of ice that flowed along the bottom of such gorges when they were above the sea.

But it appears that such erosion could be performed by heavy glaciers with the lands at their present level. A glacier three thousand feet thick would fill and press heavily on the bottom of a gorge fifteen hundred feet in depth. Therefore, should the bottom of a fiord sink hundreds of feet below the sea-level, a glacier several thousand feet thick flowing through and over it into a sea of much greater depth, the erosion at the bottom of the sunken channel would be greater than on the land above the sea, where the ice possessed less weight.

Therefore, it is not necessary that lands pierced by deep fiords should have acquired a higher level during the ice age than they now maintain. And it is probable that on the antarctic continent ice erosion may be going on at much greater depths below the sea-level than the deepest channels in the high northern latitudes. For it is likely that the temperature of a glacier is so low in such frigid regions that it holds firmly in its freezing grasp such bowlders as may become detached from the rocks, thus giving it great erosive power.

But this great eroding ability could not be maintained by glaciers in the lower latitudes, where a higher temperature would largely deprive the ice of its abrading properties except on the steep slopes of mountainous lands.

There are deposits of ice on the North American coast bordering the arctic shores, and also on Northern Siberia, that are supposed to have existed since the last frigid period, and are likely to be preserved into a future cold age, which now appears to have made considerable progress on Greenland and other ice-clad arctic shores on account of the independent circulation of the Arctic Ocean waters, which largely excludes the Gulf Stream from the polar seas; and it is for this reason that the glaciers on the elevated lands of Iceland are being enlarged and rapidly advancing. Yet, notwithstanding the gathering of ice and increasing coldness of lands largely removed from the warm Gulf currents, there are still mountain regions where glaciers may have been preserved through post-glacial times, although directly to the leeward and under the influence of the Gulf Stream and Japanese currents. These glaciers are situated in the Alpine districts of Europe and on the mountain ranges of Alaska. It would appear that, were the climate growing gradually colder in the northern temperate zones, such glaciers should be increasing in size.

Yet it is said that such is not always the case. This is probably owing to their being subject to the genial influence of the tropical currents. For, although the climate of Europe and Alaska may have been slowly growing colder for centuries, still the slow shrinkage of these once immense glaciers may still be going on, although at a much slower rate than formerly, even if the tender plants of these latitudes, because of the growing coldness, have gradually moved southward.

As to the Alpine glaciers, M. Forel reports from data he has collected that there have been several enlargements and diminutions during the last century. And since 1875 enlargements have taken place, their shrinkage being caused by warm and dry weather, while their enlargement was brought about during cold and rainy seasons. The glaciers of Alaska cannot attain much extension until the waters of the great Japanese stream acquire a lower temperature. There is at this date a small current setting down through the eastern side of Bering Strait, bearing field-ice in the spring season down to Anadyr Gulf. The Okhotsk Sea in the spring season furnishes considerable field-ice to cool the north Pacific waters, and the wintry winds which sweep down from the high lands of Northern Asia also serve to chill the Pacific seas; but all such sources of cold combined at this age have but little general effect on the vast Japanese current, which still has warmth sufficient to prevent the increase of glaciers on Alaska.

This great ocean stream in its impact against the shores of Oregon causes a high sea-level, which is mostly turned southward by the prevailing north-west winds. Still, a comparatively small stream sets along the shore of the Alaska Gulf, and also through the island passages toward a slight low sea-level, to the leeward of the Alaska peninsula; and it is probable that this current which warms these in-shore waters is favored by the difference of temperature and density between the waters abreast Oregon and the Gulf of Alaska, and it may be owing to the same cause that a small stream is sent along the eastern shore of Bering Strait into the deep portions of the Arctic Ocean. Thus because of the warm waters that proceed from the great Japanese current the glaciers of Alaska are prevented from increasing their bulk.

The only way to furnish the Japanese stream with colder water, and so cause glaciers to increase on the north-west coast of America, is through the great Humboldt current, which has its rise in the southern ocean west of Patagonia and the Cape Horn channel, where a moderate but vast high sea-level is formed on account of the great drift current of the southern ocean being somewhat obstructed on its passage through the Cape Horn channel, which is about one-third the breadth of the westerly wind-belt.

Therefore, the northern portion of the waters of the high sea-level so caused are attracted northward to the low sea-level abreast Peru, from whence they are moved by the south-east trade winds as a drift current to the equatorial latitudes, thus meeting and mingling with the returning Japanese current abreast Central America, and so giving head to the great equatorial stream which moves westward over the Pacific Ocean, partly impelled by the trade winds, and, on gaining the western side of the ocean, sends off from a moderate high sea-level a large stream to the low sea-level caused by the westerly winds abreast Japan, from whence it is drifted by the same winds over to the north-west coast of America, thus forming the great Japanese current.

Meanwhile the temperature of the Humboldt current, being governed by the temperature of the southern ocean from which it takes its rise, is cooling at a slow rate through the enlargement of ice-sheets in the antarctic regions, while the increase of glaciers on Patagonia will in time greatly add to its coolness, and so lower the temperature of the equatorial current from which the Japanese current branches, the latter current being made cooler through the increase of coldness of the former streams. Therefore, the temperature of Alaska, which is governed by the Japanese current, will slowly acquire a colder climate; and, consequently, its glaciers will increase in size sufficient to launch icebergs into the Pacific to be cur-rented southward, and so still further lower the temperature of the Eastern Pacific waters, and consequently the equatorial current from which the Japanese stream branches, and so eventually, under the above conditions, cause heavy ice-sheets to spread widely over the north-west coast of North America.

It will be seen from the above explanations how an increase of cold in the southern hemisphere is necessary to cause a wider spread of ice-sheets on lands in the northern hemisphere.

Especially is this the case to promote the gathering of glaciers on the west coast of North America. The great equatorial current while on its way to the Indian Ocean not only sends off the Japanese stream, but also the East Australian current, which is like the Japanese current, having its temperature lowered in proportion as the equatorial stream is cooled. Therefore, the southern ocean is slowly being deprived of equatorial heat from this source.

I have explained how the increasing coldness of the superior oceans of the southern hemisphere affects more or less the temperature of the Gulf Stream, which meanwhile is only able to enter a small portion of its waters into the Arctic Ocean after undergoing a long cooling process as a drift current; and, while thus mingling with the arctic waters, it is not able to prevent the gathering of ice-sheets on Greenland, where glaciers are launching bergs to float southward as far as the latitude of 40° north. Consequently, the northern seas are now being cooled as well as the seas of the southern hemisphere.

Yet this cooling process is so slow there is a lack of data to show that the temperature of the high latitudes is lowering. Our thermometrical observations are of such recent date they cannot be used to determine climatic changes which requires centuries to bring about. Still, it is generally known that the climate of Northern Europe has been accused of growing colder. The vine no longer flourishes on the shores of Bristol Channel or in Flanders or Brittany; and vineyards are no longer planted on the elevated shores of France where they flourished three hundred years ago. Arago did not refuse to believe that the laws regulating the temperature of Western Europe had notably altered. This is proved, he said, by the general retrogradation of the vineyards southward.

The recent deadly freezing of the orange groves of Florida makes it uncertain whether the cultivation of the orange can again be successful in the counties where during this generation it has been very profitable.

Travellers visiting Iceland say that the old accounts of its prosperity seem strange to those who now visit its shores; and it is narrated in the Sagas that in early times sheep could shift for themselves during winter, and that there were large forests and that corn ripened. Several years ago a correspondent of the *Spectator*, writing from Northern Russia where the Volga is locked with ice for six months in the year, stated that "the people were beginning to show increased resentment at the climate, and that there was reason to believe that the northern government of Russia would be abandoned to the desert. The people silently glide south by the tens of thousands every year, so the life of Russia was concentrating in the south."

It is now the opinion of travellers in arctic lands that the inhabitants of the Esquimaux regions are decreasing, as are also the inhabitants of Northern Siberia.

A writer in the North China *Herald*, of Shanghai, says that "the climate of Asia is becoming colder than it formerly was, and its tropical animals and plants are retreating southward at a slow rate. In the time of Confucius elephants were in use on the Yangtse River. A hundred and fifty years after this Mencius speaks of the tiger, the leopard, the rhinoceros, and the elephant as being in many parts of China.

"It is also said that the ferocious alligator, that formerly infested the rivers of South China, has retreated southward.

"The flora of the country is also affected by the increasing coldness of the climate. The bamboo is not found in the forests of North China, where it grew naturally two thousand years ago, but is still grown in Pekin, with the aid of good shelter, as a sort of garden plant only."

A letter from Hong Kong, published in the London *Standard*, reports that on the 15th of January, 1893, the temperature of Hong Kong, a tropical seaport of China, was below freezing for three days, and was colder than ever before known. The rocks and also vegetation were covered with a coating of ice. The thermometer at times stood at 23° and 26° Fahrenheit.

I have previously explained how the slow increasing coldness of the northern temperate zone is also being carried out in the southern hemisphere. The meteorological records for the lofty table lands of Ecuador, although very incomplete, furnish strong evidence to show that the mean temperature of that region is gradually lowering.

Observations made by Boussingault at Quito in 1831, compared with those from 1878 to 1881, showed a decrease from 15.2° Centigrade to 13.27° Centigrade.

Records made by Hall from 1825 to 1827 give averages of 16.1° Centigrade, 15.52° Centigrade, and 15.6° Centigrade. This decrease holds good for all points in the inter-Andean region where records have been kept.

Yet we know that the falling temperature in the northern temperate latitudes is not brought about by a yearly increase of cold, because, when the arctic channels are somewhat obstructed with icebergs, the movement of arctic waters through them is lessened; and, therefore, during such times the Gulf Stream, meeting with less opposition from arctic currents while flowing northward, is able to move a larger volume of its waters into the arctic seas, thus warming their waters sufficiently in a few seasons to clear the obstructed channels, and also somewhat soften for several successive years the temperature of such lands as border on the seas of that region.

And in this way we account for the mild seasons which at times follow those of lower temperature in high northern latitudes.

But, when the detained icebergs are set adrift, and currented into the temperate North Atlantic, the heat consumed while melting such numerous bodies of ice is able to more than overcome the warmth gained during the temporary detention of ice in the northern seas. Thus, under such considerations, it appears that the conditions are favorable for the growth of glaciers in the high northern latitudes.

I have pointed out the manner in which the superior oceans in the southern hemisphere are obtaining a lower temperature, and how they impart their coldness to the tropical currents, and in this way slowly cool the waters of all oceans. Thus it appears that the northern temperate zone, with all other parts of the earth, is slowly approaching a cold epoch.

Several writers on climatic changes have expressed their views as to the number of glacial and mild periods that have been perfected since the conditions have been favorable for their appearance on the globe. According to my views, while considering the reasons for the occurrence of the great glacial periods which have left such extensive traces on the land, it seems certain that two very cold epochs have possessed the earth, separated by a warm period; and, possibly, other preceding cold epochs of less intensity have possessed the high latitudes, with intervening periods of mildness. But the earlier cold periods, if they ever existed, were comparatively short, because the Cape Horn channel during such times possessed less capacity than in the later periods, and, therefore, was more easily and quickly obstructed by the natural methods previously explained.

Consequently, the independent circulation of the southern ocean was sooner arrested than during the later epoch, when the channel had become enlarged by erosion from heavy glaciers and icebergs; and meanwhile the same conditions may have governed the arctic channels which give an independent circulation to the arctic waters which surround Greenland, and thus, in connection with cold epochs in the southern hemisphere, have caused periods of cold of small intensity to occur in the high northern latitudes, and it may happen in the future that more ice periods will be perfected than the one now progressing.

Still, it is well to bear in mind that the Cape Horn channel, which is the real cause of glacial periods having occurred in both the northern and southern hemispheres, in the manner previously explained, is being made wider and deeper during each succeeding ice age. For this reason the latest cold epoch will require a longer continuance of cold to obstruct the channel than the cold period preceding. Therefore, it appears that the time will come when there will be such great accumulations of ice stored on the land and in the sea before the enlarged Cape Horn channel can be closed that, when it is closed, there will not be sufficient warmth remaining in the tropical seas to unite with the sun's rays to subdue the intense cold stored in the immense gatherings of ice. And thus the earth, which began its career with a warm temperature, and so continued for long ages, will finally terminate in an endless glacial age.

The statements made by General Cowell in *Science* of Nov. 25, 1892, in reference to the alleged discovery of the second rotation of the earth by Major-general Drayson,. represents the discovery as affording a new solution for the cause or causes of an ice age.

The second rotation as defined consists in the pole of the heavens describing a circle around a point which is ascertained to be situated six degrees distant from the pole of the ecliptic. And it is asserted that by a knowledge of the second rotation it is proved that a variation of twelve degrees in the extent of the arctic circle and the tropics occurred not later than 13,500 B.C., "the tropics varying in distance from the equator from the minimum of 23° 25' 47" to the maximum of 35° 25' 47", thus extending the torrid zone during its widest expansion from Cape Hatteras to the river Plate.... It is calculated that at this date we are about 403 years distant from the time when the pole of the heavens in its revolution, the pole of the ecliptic and that of the second rotation, will be in the same colure,—that is, in the year

2,295 A.D.; and then the least differences in temperature between summer and winter will be experienced. From that time forward this difference will increase, and about 6,000 years later, or about the year 8,300 A.D., the earth will enter the next glacial period, and attain its greatest severity about the year 18,136 of our era." General Cowell does not state how the widening of the tropical zone, as above set forth, would bring about a glacial period. The winters of the temperate zones would evidently be colder than now; but, on the other hand, the summers would be proportionally warmer, while the westerly winds above the latitudes of 40° would prevail the same as now.

Therefore, their general effect on the surface waters of the ocean in the high latitudes would not be changed with such an extension of the tropical zone, neither would the trade winds change their general direction with a wider torrid zone; yet the boundaries of the trade winds and also the westerly winds would be more shifting according to the declination of the sun, such winds being governed as now by the position of the sun during the summer and winter solstice. Yet the natural process for moving tropical water into the high latitudes, or excluding it therefrom, would not be greatly changed.

Consequently, the expansion of the torrid zone to the latitudes named by General Drayson would not affect the climate of the hemispheres sufficiently to cause a frigid epoch. On the contrary, the summer monsoons, which now blow from the north-east, along the shores of Eastern Africa, and also along the coast of Southern Brazil, would be much stronger with a vertical sun in midsummer as far south as river Plate, thus forcing the surface waters of the tropical oceans into the higher latitudes with greater facility than at this age.

Moreover, according to the statements of General Cowell, the present period of mildness should be on the increase, and obtain perfection in the year 2,295, or about 400 years hence; while, on the contrary, according to the explanations we have given in the preceding pages, there is much to show that an ice age is advancing, and has made considerable progress in the high latitudes of both hemispheres. Furthermore, if the second rotation, as claimed by General Cowell, is able to perfect a glacial period at regular intervals of 31,600 years, it seems that traces of frigid epochs should not be confined to late geological records, as there appear to be little or no traces of glacial work prior to the Quaternary or Post-tertiary periods.

It appears that explanations so far given, which depend on the astronomical theory to account for the ice age, are not in harmony with well-known geographical facts. The explainers neglect the attention due to the great prevailing winds which since the earlier geological ages have, in connection with continents, moved the surface waters of the ocean from torrid latitudes to colder zones, and from the colder zones to the warmer latitudes.

This exchange of ocean waters between the zones is as old as the continents which shape their courses. The important change wrought in the ocean currents sufficient to have caused the glacial age which ended the early warm epochs was brought about through the action of the prevailing winds, which, in connection with the form of continents, became able to move the ocean waters from the northern hemisphere into the southern sufficient to submerge the low lands of the southern hemisphere, causing a great diversion of the tropical currents from the high southern latitudes, such as I have pointed out in preceding chapters.

Those writers who believe that ocean currents have been the cause of great climatic changes have suggested that the existence of an ancient channel through the isthmus of Panama would have caused a frigid period on lands bordering on the northern shores of the Atlantic by turning the head-waters of the Gulf Stream into the Pacific Ocean.

Professor Agassiz thinks that such a channel existed during some remote geological age, judging from the semblance of the fauna pertaining to the Caribbean Sea and Pacific Ocean.

Yet it may be said that an open channel through Central America would have connected two high sea-levels.

For this reason there would be little or no exchange of water between the Caribbean Sea and the Pacific Ocean.

The high sea-level on the Pacific side is caused by the prevailing north-west winds which blow down the North American coast past California as far south as Central America; while, on the other hand, the south-east trade winds impel the surface waters of the South Pacific along the coast of Peru down to the equator, and so onward 5° to 8° north latitude. Thus the space between the ending of the two ocean winds obtains a high sea-level, corresponding to the high level of the Caribbean Sea. This has been proved from levellings for the Nicaragua ship canal.

Consequently, the Atlantic waters would not run into the Pacific Ocean, even if a channel opened through Central America.

Therefore, the Gulf Stream has never been turned away from the North Atlantic.

Writers, while seeking a cause for the mild climate of ages preceding the glacial epochs, have thought that during such times channels opening through Asia from the Indian Ocean by the way of the Persian Gulf into the arctic seas would be the means of furnishing the Arctic Ocean with warm water. But it is evident that such a movement of water could not be brought about, because the winds would not be favorable for it. For, when we reflect that the prevailing winds would blow in the same direction as now, and that the seas of Eastern Europe and Western Asia were enlarged during the warm epochs, it seems that they would obtain high levels superior to the high level seas of the Indian Ocean.

Besides, we should consider that there is a continuous range of high land separating the Persian Gulf from the northern seas, which probably existed anterior to the ice age. Still, during later periods, while the ice-sheets were being melted from the northern hemisphere and also on the ending of the last ice age, the Isthmus of Suez was submerged, as were all other low lands in that latitude; but it is probable that the waters of the high sea-level of the Indian Ocean abreast tropical Africa did not flow largely into the Mediterranean Sea for the reason that the enlarged

European seas, being within the westerly wind-belt, maintained a high sea-level, while at the same time the high level tropical Indian Ocean waters were strongly attracted into the southern oceans through the Mozambique and Agulhas currents in the manner I have previously explained. Yet the waters of the high sea-level of the southern European seas must have been strongly attracted to the low sea-level abreast the Canary Islands.

While considering the causes which brought about the glacial periods, it is well to reflect that the natural mode of action which could have produced a frigid age was as extensive as the surface of the globe; and, therefore, any geographical change that would affect only a comparatively small portion of the earth cannot serve to account for ages of warmth which extended over the globe, or for glacial epochs which were separated by warm periods of time, which seem to have affected all lands and seas.

And it appears from the geographical explanations given in preceding pages of the general movements of the winds and currents of the sea how impossible it is for heat to be conveyed to the antarctic latitudes sufficient to prevent the growth of glaciers on their lands while the Cape Horn channel is in possession of its present capacity.

For, as has been shown, this channel furnishes opportunity for the westerly winds to impel the surface waters of the great southern ocean constantly around the globe, and so largely turns away the tropical currents from the high southern latitudes.

Consequently, there seems to be no method yet devised through nature's mode of action that can carry sufficient heat into the antarctic latitudes to melt the ice-sheets from the southern continent, or even arrest their growth, while the Cape Horn channel maintains its present width and depth.

Therefore, the increase of glaciers and icebergs will slowly continue until a glacial epoch is perfected.

And it seems that this arrangement for bringing about a frigid age made slower progress in its early stage than at this date, owing to there having been a lack of glacial ice in the polar regions to produce icebergs for cooling the ocean waters. But the independent circulation of the great southern ocean, after turning away the tropical currents from the high southern latitudes for thousands of years, did at length cause glaciers to form on the antarctic lands, which have been slowly, but constantly increasing; and, consequently, the cooling of the ocean has been accelerated proportionate to the increase of ice-sheets. Therefore, with the cooling process so well advanced as it now appears to be, it seems that more than half of the time required to bring a frigid age to perfection has been expended since ice-sheets began to gather on the antarctic shores. For, when we realise how the facilities for making ice have advanced through the increase of glaciers in both hemispheres, and how large a portion of the ocean waters have been cooled below a temperate or tropical temperature even in the torrid latitudes where the warm upper waters of the ocean have been reduced to a comparatively thin stratum when compared to the vast bulk of the cooled under-waters, it appears that the cold will increase at a faster rate for the next thousand years than was the case during the last ten centuries. Therefore, the climate will be less favorable for plants and animals existing on lands in the high latitudes for the next thousand years than during the ten centuries preceding; and, when we take into consideration the accelerative growth of a frigid epoch, it seems that the increasing cold will in a few thousand years drive the greater portion of both plants and animals from the now temperate latitudes to maintain an existence in the tropical zone, where a large part of the existing species of such life must have taken refuge during the last ice period.

And, from what can be learned from the relics of man's prehistoric life, it seems to point to the lands of the tropical latitudes as having been his home during the frigid ages; and, because of his long undisturbed residence in favored portions of the tropics, he there attained his earliest civilization. For it appears that the tropical zone was not only less burdened with ice in glacial times than the higher latitudes of the globe, but was also more exempt from the great flooding of lands which obtained in the more northern latitudes through the shifting of the ocean waters, from causes set forth in the preceding pages. Yet it may be said that the low lands of the tropical zone south of the equator during cold epochs were much more extensive than at this age, on account of the shrinkage of the sea, because of the great amount of water evaporated from its surface, and stored in ice-sheets on the great continents and islands. Hence the reefs and shallows which surround such tropical islands as include the Seychelles Archipelago, and also the extensive banks covered with shoal water in that portion of the Indian Ocean, were during the glacial period elevated above the surface of the sea, possessing a climate favorable for vegetable and animal life. But, owing to the great rain-fall of that region, it is probable that the highest lands were glaciated, as it is reported that granite bowlders still rest on the mountain slopes of the highest island. The numerous islands and shoals of the south-western tropical Pacific must also have afforded wide land areas, with a temperate climate, owing to their having been situated on one of the warmest regions of the earth during the ice age.

Moreover, it is probable that these tropical lands afforded space for numerous lagoons which had little connection with the surrounding oceans, and consequently were able to maintain, in their secluded shallow basins, a warmer temperature than obtained in the open seas; and at the same time, owing to the great rainfall in such tropical portions of the Indian and Pacific regions, the waters of the lagoons were rendered less salt than the briny depths of the shrunken oceans of a cold period. Hence because of such conditions the fauna of the tropical seas were preserved from the destructive rigor which beset the earth during the frigid epochs.

Lector House believes that a society develops through a two-fold approach of continuous learning and adaptation, which is derived from the study of classic literary works spread across the historic timeline of literature records. Therefore, we aim at reviving, repairing and redeveloping all those inaccessible or damaged but historically as well as culturally important literature across subjects so that the future generations may have an opportunity to study and learn from past works to embark upon a journey of creating a better future.

This book is a result of an effort made by Lector House towards making a contribution to the preservation and repair of original ancient works which might hold historical significance to the approach of continuous learning across subjects.

HAPPY READING & LEARNING!

LECTOR HOUSE LLP
E-MAIL: lectorpublishing@gmail.com